Landlord William Scully

wswswswswswsws

HOMER E. SOCOLOFSKY

wswswswswswswswswswswswswswswswswswswswswswswswswswswswswswswswswswswswswswswswswswswswswswswswsws

Landlord William Scully

wswswswswswswswswswswswswswswswswswswswswswswswswswswswswswswswswswswswswswswswswswswswswswswswsws

THE REGENTS PRESS OF KANSAS
Lawrence

Printed in the United States of America

Library of Congress Cataloging in Publication Data

Socolofsky, Homer Edward, 1922-
Landlord William Scully.

Bibliography: p.
Includes index.
1. Scully, William, 1821-1906. 2. Businessmen—Middle West—Biography. 3. Landlord and tenant—Middle West—History. I. Title.
HD210.M53S386 333.5′4′0924 [B] 78-31477
ISBN 0-7006-0189-9

TO

Penny

Contents

wswswswswswsws

List of Illustrations

WSWSWSWSWSWSWS

Preface

wswswswswswsws

As a landlord in the American midwest, William Scully had an unparalleled role, almost from the time he purchased large quantities of low-cost land in Illinois in 1850 to his death in 1906. Frequently mentioned in studies of that period as "the object of as much ill feeling and political agitation" as any other frontier landlord in the entire country, William Scully led a life that had profoundly important parts. Never before has his entire life been presented in such a thorough manner as to provide the interrelationship between his years in Ireland, where he gained national notoriety, and his career in the United States.

In fourteen counties in the states of Illinois, Kansas, Missouri, and Nebraska and in the Irish counties of Tipperary and Kilkenny, the name of William Scully is well known. The unique land system that he used on his 225,000 acres of American agricultural land, farmed by some fifteen hundred tenants at the time of his death, attracted attention to this "most extensive American landowner."

William Scully possessed an extraordinary determination to succeed as a landlord. Born into an Irish landed family of moderate wealth in 1821, he inherited about a thousand acres in 1843. With his savings he went to Illinois in 1850 and bought thousands of acres of cheap government land, thus setting the pattern for later years. In the 1860s, after his return to Ireland, his interpretation of the landlord's legal position pushed tenants to violence, and he almost lost his life. By that time his wealth had grown so that he bought land in other American states and established a large number of estates that could be supervised by agents, with slight attention required from the landlord.

Hostility toward William Scully was rarely expressed by American tenants; it came instead from nearby newspapers and politicians who viewed Scully's creation as un-American. Laws passed to restrain ownership of land by nonresident aliens were directed at what Scully had done. Not acting the way he had in Ireland, Scully made no public response to attacks on him, and American attitudes toward him in later years were to change. Some reports were unsure about how to treat him, and others held him up as an early conservationist and as an organizer of a sizable estate that would be held together by his descendants. More than seventy years after his death, almost all of the Scully estates remain in the hands of grandchildren and great-grandchildren. Present-day procedures for leasing land to tenants follow the pattern set down by Scully as his extensive holdings gave him recognition as the American landlord with the most farms.

The intent of this book is to narrate for the general reader, as carefully and completely as possible, the life of William Scully. The lives of thousands of tenants and their families have been profoundly influenced by the Scully system of landlordism. This study was made primarily for them. Some may draw on it to examine leasing systems used by other landlords with large holdings to make comparisons between their practices and those of Scully. I did not intend to analyze Scully's leases, the tenure of his tenants, the income of the landlord and tenant, or the social and economic role of Scully's landlordism in comparison with that of other landlords.

I hope that this book will provide another dimension in the enjoyable study of American history, its relation to old-world patterns and to economic developments. If that purpose is served, I am fully repaid.

I have been accumulating material for this book for more than thirty years, therefore I cannot possibly provide a complete list of all those who gave help and encouragement. Thanks is due to colleagues Verne Sweedlun, who directed my initial Scully research; Bower Sageser, James Carey, and Robin Higham, all of Kansas State University; and George M. Curtis, III, of the College of William and Mary, for extensive comments on portions of the manuscript. Faculty research grants from Kansas State University helped this study along the way. Paul W. Gates, Paul J. Beaver, and Russell L. Berry helped by working other portions of the Scully story which found application here. Margaret Beattie Bogue of the University of Wisconsin, Madison, gave a needed critique on a shorter version of this study, and Mark Plummer, Illinois State University, located an important letter. Agents for the Scully estates—James M. Stewart of Lincoln, Illinois; J. M.

Quackenbush of Beatrice, Nebraska; and D. W. Montgomery of Marion, Kansas—made valued contributions by providing access to Scully records. Mrs. Violet Scully helped by providing needed insight, and the four grandsons of William Scully—William, Robin, Michael, and Peter—were generous with their time.

Nyle H. Miller of the *Kansas Historical Quarterly*, Lorrin L. Morrison of the *Journal of the West*, and James H. Shideler of *Agricultural History* published portions of this account. Nedra Sylvis's typing speeded the preparation of the final manuscript. Inestimable has been the support and help of my wife, Helen Wright Socolofsky. Nothing I have written, however, may be blamed on any of these friends. In the final analysis, this biography is my version of an extremely complicated and interesting person.

wswswswswswswswswswswswswswswswswswswswswswswswswswswswswswswswswswswswswswswswswswswswswswswswswswswswsws

1

Privilege and Tragedy: The Early Years

wswswswswswswswswswswswswswswswswswswswswswswswswswswswswswswswswswswswswswswswswswswswswswswswswswswswsws

"FEARFUL AGRARIAN OUTRAGE IN TIPPERARY. ATTACK ON WILLIAM SCULLY, Esq." These startling words in the small-type headlines of 1868 alerted Irish and English readers to the "terrible affair at Ballycohey" on August 14, when landlord Scully and two policemen were wounded, and Scully's agent and another constable were killed. Having received two shots in the jaw, one in the neck, and others in his body, Scully was not expected to recover.[1] A century later, in a cemetery near Ballycohey, a monument was raised to memorialize "The Fight against Landlordism" which "So Frustrated the Despotism of Alien Landlords."[2] In the usual sense, neither William Scully nor his family could be classified as alien.

In William Scully's case, the struggle had begun less than eight miles away, in Kilfeacle House in county Tipperary, on November 23, 1821, when he was born to Denys Scully and his wife, Catherine Eyre Scully. For four hundred years the Scullys had been a prominent landholding Irish-Catholic family in British-dominated Ireland. This fifth son and ninth child, christened William Francis John, had such a lowly ranking in a large family that his future position in Ireland, even among the lesser gentry, seemed unassured.[3]

Ireland, for many years, had been controlled by a Protestant English government. Land was owned, for the most part, by English nobles who were loyal to the crown and to the Church of England. Blood lines meant everything to this group of nobles. These Anglo-Saxons felt that they were superior to the Celtic races, an attitude that was bitterly resented by most of the Irish peasants who were Catholic. They served in an inferior role as servants, farmers, and herdsmen for the landlord class. Members of this peasant class, who had the lowest income in the land, had few privileges, and their burdens were heavy. Long on memory, they recalled an earlier, brighter time, when clan chieftains, their own ancestors, owned and controlled the Emerald Isle. To them it seemed almost like yesterday, and it changed nothing in their minds that for hundreds of years, English landlords and their families had been born in Ireland, had lived there, and had died there. To the steadfast Irish the English were aliens, not only because of the violent overthrow of earlier Irish governments and because of their ownership of Irish land, but because of differences in religious commitment. A belief persisted among Irish peasants that "ancient Irish families would recover their forfeited estates."[4]

The Scully family was descended from the O'Scolaidhe clan, an Irish sept which originally was located in Westmeath. During the twelfth century, pressure from Anglo-Norman invaders pushed most of the family into county Tipperary and elsewhere.[5] Denys could easily trace his lineage back six generations, to James Scully, born April 12, 1571, in King's County.[6] After the Restoration of Charles II in the seventeenth century, members of the family again settled in county Tipperary near Cashel. There, in the "Golden Vein," they occupied some of the most fertile soil in all of Ireland.

The Kilfeacle location has a long recorded history. Behind the Kilfeacle church, on the road between Tipperary and Golden, was a mote or hill on which an English castle had been erected in the late twelfth century. Various governors of Ireland had lived there as late as the sixteenth century. Later the castle was destroyed, and all brass and iron was removed.[7] The Kilfeacle manor house, in the barony of Clanwilliam, was about a half mile north, behind high stone walls erected to the west of a stream that flowed through the valley. A further half mile north were ruins of an ancient stronghold. Round about in the bright-green grasslands were many hovels and homes of other families. In the late seventeenth century, taxes were paid on twenty-one hearths in Kilfeacle town, an indication of the number of dwellings at the time.[8]

To have Irish catholic families such as the Scullys continuing as landlords and members of the lesser gentry was unusual and unlikely. But the pragmatic English permitted such arrangements when the advantage was on their side. No doubt, earlier leaders of the family served the British in time of war. As gentry they were active in local government and were dependable allies to legal authority. Finally, when the British relaxed regulations that had been designed to hold down Irish Catholics, there were members of the familywh erved in national governmental positions.

Denys Scully, William's father, received many privileges not generally granted to his countrymen or, for that matter, given to earlier generations of Irish Catholics. Denys's father, James, who had extensive landholdings, had his residence also at Kilfeacle. There Denys was born in 1773, the second son in a large family. He received a splendid education for the time and was permitted in 1794 to enroll in Trinity College at Cambridge University. Denys was either the first or second Catholic student at Trinity in about two hundred years. But he was not allowed to graduate, because one stipulation for graduation was that he conform to Britain's established church, a step that he would not take. In 1796 he was back in Dublin, where he was admitted to the bar, virtually the only profession then open to Irish Catholics.

By the time that Denys Scully married Mary Huddleston in 1801, he had become heir apparent to his father's position, as his older brother had died. He was slow of speech, but he made himself known by writing several pamphlets supporting the government. Early in 1805, possibly after the death of his wife, Denys was a member of an Irish emancipation deputation which unsuccessfully sought the backing of William Pitt the Younger. By this time Denys was maintaining a Dublin residence on fashionable Merrion Square, across the park from a house that was later occupied by Daniel O'Connell, whom he joined in an effort to obtain greater rights for Irishmen. At that time the character of his writing changed diametrically from its previous progovernmental stance.[9]

Dublin, long the leading cultural and political center of Ireland, had fallen to non-Irish control during the Norman invasion of the twelfth century. A garrisoned Anglo-Norman enclave was organized around Dublin two centuries later. High stone walls, masking trees and huge estates, screened the entrances to the city that came to dominate the Irish scene. In time, colleges and universities, chapels and cathedrals, government buildings, and a business center graced the banks of the river Liffey where it emptied into Dublin Bay. The English lord lieutenant lived nearby in a large mansion on grounds that were later incorporated into Phoenix Park. His seat of governance was Dublin Castle, and his rule, to the native Irish, was that of a foreign master. Social and cultural activity in Dublin was a mixture possessing both Irish and English precedents. Irish gentry associated with their English counterparts, while members of the Irish lower classes and a small middle class had little opportunity for social contact with gentry, either native or foreign. Many landed Irish families maintained town houses in Dublin, and some Irish landlords spent only a small part of each year on their landed estates. The social setting in Dublin, for many of the Irish gentry families, provided timely intimate contact with others in this small, select upper class, an opportunity to form alliances, and a means of solidifying their position.

The marriage of Denys Scully to Catherine Eyre in 1808 brought closer ties with English Catholics, as her home had been at Highfield and Newholt, Derbyshire. The contractual marriage settlement of September 7, 1808, gives an idea of Denys Scully's existing fortune in describing the lands and tenements of seven farms, totaling almost sixteen hundred acres, from which he received an annual rental income of more than twelve hundred pounds. Farm names such as Ballinaclough, Knockroe, Folcherstown, Rathmacan, Gortnagap, Keil Ballyonstra, and Springmount, located in the baronies of Clanwilliam, Eliogarty, and Craunagh and in the counties of Tipperary and Kilkenny, show ancentral ties to old traditions, and they alert one to the unique and complex character of Irish landholding rights.[10] Denys received other lands when his father died in 1817.

Following his marriage to Catherine, Denys Scully began active and covert agitation to gain some measure of justice for Catholic views in Ireland.[11] By 1812 he had prepared a manuscript, which was published anonymously as a small booklet with the lengthy title *A Statement of the Penal Laws, which Aggrieve the Catholics of Ireland: With Commentaries.* Legal sanctions were harshest for publishing such materials; therefore its Dublin printer, Hugh Fitzpatrick, was heavily fined and imprisoned for eighteen months.[12] Since the authorship of the disruptive pamphlet was not immediately discovered, no harm came to its author. When it became general knowledge that Denys Scully was responsible for this piercing attack on English authority, his popularity among his Irish countrymen was assured. For a while there was no reprisal from the government, but the opportunity for harassment came in 1818, when Denys's father died and the estate was probated before an English-organized court. Denys's friends and his immediate family were certain that the resulting family lawsuit was instigated by the government because of his harsh criticism of English authority. The court fight was so bitter, costly, and time-consuming that Denys withdrew from public life. By the time the case was settled in his favor, his health was so enfeebled that he could not resume his public career.[13]

This was the life into which William Scully was born, the youngest child in a well-established landed family. His venerated grandfather had died three years before William was born. He barely got to know his own father, who was unwell and died before William's ninth birthday. The burial of his father among family graves around the ruins of an ancient cathedral on prominent, three-hundred-foot-high Rock Cashel gave a special position to this Irish family and must have made a strong impression on the mind of the young William. Later, one of William's brothers raised a mausoleum, with an elegantly carved Celtic cross, over his father's grave.

William's eldest brother, James Vincent, was twelve years his senior. As first inheritor of his father's position, James received a superb education: like his father, he attended Trinity College, Cambridge; and at the age of

Rock Cashel, county Tipperary, Ireland, the site of several ancient churches and the Scully mausoleum (shown on the high point of the rock, to the left of the ruins), erected in 1867 in memory of Denys Scully. Members of the Scully family are entitled to burial in this cemetery.

twenty-one he was admitted to the bar at Gray's Inn. By that time a major share of the family's estates were under his direction. He was fully aware of the advantages possessed by the landlord class, almost to the state of arrogance, and he was admired by his youngest brother.

Next in line was Vincent James, who was eleven years older than William. Less was expected of a second son in a landed family, although eventually that was to change for Vincent. Denys and Catherine's third child was another son, Rodolph Henry; and then came three daughters, Catherine Julia Mary, Mary Anne, and Juliana who died before reaching her first birthday. Thomas Joseph Denis Aloysius was four years older than William, and a second sister Juliana was two years older.

In this active upper-class family, William was the baby, the one to be coddled. Little was expected of him, and only a small portion of the family's claim to wealth would normally reach him, for all of his older brothers had a prior right to the father's political position. Each child, however, might be granted land in a will. With the girls in the Scully family it was different. Provision was made for them to receive annual stipends from the landholding members of the family, and when they married or went into a religious order, a dowry was provided.

Compared to his later years, this period of infancy and childhood for William was tranquil. There were many members of the immediate family, and there were numerous servants in the Scully household. The two family dwellings, at Kilfeacle and on Merrion Square in Dublin, were ninety miles apart, or two hard days by coach. Major portions of the family spent long periods in the Dublin house. Travel back and forth between Tipperary and Dublin was arduous and time-consuming before the railroad was built. Then, the nearby stations of Limerick Junction and Dundrum provided easy access to faster, more-comfortable trains. Occasionally, family members would travel outside Ireland, for education, rest, or recreation, and visitors flocked to the Scully manor house in Kilfeacle or to the town house, no. 13 on the south side of Merrion Square. There Catherine Scully, surrounded by silver and fine linens, was a gracious hostess for a dozen years following her husband's death.

Typically, other things occupied the time and attention of the children. There were the procession of the seasons, the growth of crops and farm animals, play with brothers and sisters and with cousins and other children. Rudimentary learning of letters, arithmetic, and the Catholic catechism occupied many hours in the childhood years of upper-class children. Special days in the church year and days of recognition for various family members were celebrated.

For growing boys there was the Irishman's delight in a good horse and equestrian ability, and these were fully a part of the yearning and learning of the Scully brothers. One of the privileges of their class was to ride to the hounds in a fox hunt, and it was a jealously guarded prerogative. It made no difference if the fox sought an escape through a peasant's field and if hounds and horses tramped the crops down in pursuit. To soar over stone fences on a fine hunter was an ultimate thrill that was particularly exhilarating to several of William's brothers. Two hunt clubs operated in the vicinity of Kilfeacle House, and in wintertime, hunts were held two or three times each week.

Another privilege of the landholding gentry was the right to possess firearms and to shoot birds and wild game. Ownership of firearms was forbidden to the lower classes, and harsh penalties were meted out to those found guilty of poaching on the game preserves of the upper classes. Peasants who were caught carrying firearms were severely punished. William and his brothers enjoyed hunting ducks on the pond below Kilfeacle House or shooting rabbits and other game in nearby fields. The possession of expensive shotguns and rifles and the ability to use them were prized in the Scully family.

William became a fine horseman, but he was not captivated by such skill. He was more intent upon preparing for a future which in all likelihood would be bereft of a sizable inheritance. No doubt there were family discussions, which he did not fully comprehend at the time, about the operations of the Scully's bank at Tipperary, one of the few banks in Ireland to survive the financial crises of 1820 and 1825. Probably another topic of conversation was the fact that the bank came to be controlled by John Sadleir, a relative, and then it had a scandalous downfall.

The Scully Cross on the family mausoleum on Rock Cashel.

The family's relationship with its tenants remained the most important business concern. Even in the 1820s county Tipperary was gaining a reputation in all of Ireland as the most likely area where Protestant landlords were an anathema. Irish peasants were quick to accept the suggestion that the hated Protestant landlords were "believers of a false religion who cannot escape perdition" and that they were "robbers of former Catholic landlords."[14] Violence was directed toward landlords; their animals were mysteriously injured, poisoned, or killed; and their servants were terrorized. Arson was a perennial problem. Even Catholic landlords were bound to receive threats when agitation built up between tenants and landlords. Some Catholic landlords suffered bodily injury and even death from unknown assailants.

Like his older brothers, William received a basic education at Kilfeacle

House, with some preparatory training in Dublin. Yet, while his older brother James attended Trinity College, Cambridge, and Vincent went to both Trinity College in Dublin and to Trinity College, Cambridge, William had to acquire his formal education in another, less-prestigious way. As a gangling youth, two months shy of his fifteenth birthday and approaching his mature height of six feet and two inches, he was sent to the Catholic-supported Stonyhurst College, which was operated by the Society of Jesus, near Blackburn in Lancashire (England).

As a Jesuit college, Stonyhurst could trace its Continental beginnings to 1593. Removal to Lancashire came in 1794, at the time of the French Revolution. In England it barely existed for many years because of official opposition from members of the Catholic hierarchy. Finally, in 1829, the pope sanctioned the development of Stonyhurst and other Catholic institutions in England.

To get to Stonyhurst it was necessary for William to cross the Irish Sea from Dublin to Liverpool, and then to travel some thirty miles to the northeast. There he spent a single academic year, in company with more than a hundred other boys, from September, 1836, to July, 1837. Classical education was emphasized, much the same as in other public schools in England. Because of his short stay, Stonyhurst made only a slight impression on William, and he, in turn, left little mark on this venerable college. Years later, William would say that he "had learned the economics of agriculture in schools," perhaps a reference in part to his sojourn at Stonyhurst.[15]

After his year at Stonyhurst, William returned home. In the fall of 1837 he began training for a law career in the usual pattern of that time. He was apprenticed in a solicitor's office in Dublin to read law and to perform errands. Although he did not enter the law, William's several years of reading it in Dublin, under the expert guidance of an established solicitor, was an excellent preparation for the time when he began to accumulate land in the 1840s. In the meantime, James Scully was having increasing difficulties with tenants on his extensive estates, and his admiring youngest brother, William, was watching him closely. In many ways James and William were alike.

When Denys Scully died in late 1830, James inherited most of the land. He had just turned twenty-one years of age the previous month, and he was admirably equipped to be active in carrying out the landlord's duties, being college trained and already a member of the bar. James was sure of himself in his new role. A tall, powerful man, he possessed the courage of his convictions as he took over extensive properties, including Kilfeacle House.

In keeping with his landed position and his rank as a gentleman or esquire, James Scully became one of the magistrates for the South Riding of county Tipperary. This important position in local government was shared by other leading gentry in the area; and because of it, Scully also served as a grand juror in cases called before the local judicial inquest or a court of assize. Politically, Scully was described as a "thorough uncompromising Reformer," indicating his opposition to the manner in which the Corn Laws were carried out, which was believed to be responsible for the impoverishment of Ireland. He also was, "and ever had been, a Repealer," who like his father before him was unalterably opposed to the Act of Union of 1801, which had abolished the Irish Parliament and merged the kingdoms of Ireland and England.[16]

One of James Scully's first actions as a landlord was to examine the rent rolls and to make minor adjustments, mostly downward. Certain tenants received a reduction of ten shillings per acre, and several tenants who were in arrears for two or three years' rent had their obligations canceled. Such tenants thought that James Scully was a good and kind landlord, and they responded with some affection and honest industry on their small holdings. However, tenants who had a reputation for idleness or disorderly behavior found landlord Scully an uncompromising master who was willing to go to great lengths to force them from his estates. He had no qualms about using every legal means, which provided immense advantages to landlords, in evicting troublesome tenants. Eviction, even though it removed the unwilling or lazy tenant from a mere existence on a piece of land that was much too small to provide an adequate return, was synonymous to banishment for most tenants. They would be forced to leave the only place known to them—their home and relations. Most tenants on Scully's estates were from families who had worked the same land for generations.

During the 1830s, agrarian outrages—the cryptic name for violent attacks on landlords—were more widely reported in the Irish press and in police records. Types of outrages ranged from broken windows, housebreaking, arson, threatening notices, and injury to or theft of property, on the one hand, to assault, robbery of arms, firing into dwellings, aggravated assault, and homicide, on the other. Landlords, their families, and those who served them were the objects of attack; and illegal groups such as the "Whiteboys" and the "Ribbonmen" organized much of their opposition. What the poorest classes particularly opposed was the landlords' effort to enlarge individual tenant holdings at the expense of smaller tenants. One example of violence occurred in 1838, when unknown attackers shot two land agents as they were passing by Ballinaclough, one of James Scully's estates.[17]

Other outside forces, such as the movement for enclosures in England and the development of new occupations in English mines and factories, which brought a demand for agricultural products from a growing urban market, were influencing Irish response. Given all political, religious, and cultural factors, in addition to strong reluctance from Irish peasants to change from subsistence farming to staple agriculture, reform and modernization faced insurmountable problems. Rural Ireland, even before the famine, was a time bomb. The inevitable explosion would produce no winners.

Increasingly hard times for farmers in Ireland, in part due to changing tariff regulations of the early 1840s, pressed hardest on tenants with high or rising rents. James Scully's tenants were gravely affected by the economic slump. To counter this, Scully, who normally employed about forty people around the manor house and on his various estates, in the summer of 1842 increased the number of his staff to more than eighty, including boys and girls as young as eleven years of age.[18]

Usually, Irish tenants owned their dwellings and other improvements, while the landlord owned the land. Somewhat earlier, one of Scully's tenants sold his improvements and his lease to a new tenant for eighty pounds. He planned to use the money to emigrate to America, but then he changed his mind. He sought return of his lease for the eighty pounds, but was unwilling to pay anything extra for crops in the ground. The new tenant was opposed to a resale under those circumstances, and he appealed to Scully, who sided with him.[19]

By that time a number of contemporary police accounts say that James Scully had become "a severe and harsh landlord" and a grinder of the poor. From some tenants he was exacting a "rack-rent," or very near the full annual value of the land they farmed.[20] In a police report of 1842, James Scully's conduct as a landlord was described as being "at variance with a number of people to whom he set what is termed conacre at a rent of £14 per acre, half the money paid in advance. He would not allow the removal of the potatoes until he was paid the other half." When these renters asked for "an abatement on the sum of the back rents of the crop," Scully objected sternly, almost to the point of violence.[21]

Nevertheless, reports out of that large Irish county in late 1842 claimed that there was "Tranquility for Tipperary." The "peace & good order" of the county, which had been so unruly in the past, was attributed to the Tory government then in power.[22]

In spite of these news stories, there was no tranquility for James Scully. Anonymous threats were delivered surreptitiously to the door or pinned on the front gate of Kilfeacle House, but the landlord was unmoved—he knew his rights. As he had never married, he had only himself to protect, and he felt adequate to the task. He failed to understand the impact of his rental policies on his tenants, and when his family and friends suggested that he reduce his rentals as a protection to his own life, he refused to listen. He knew that English and Irish law gave tremendous authority to the owner of the land, and he was sure of his own ability to handle whatever would come his way. About his only concession for his safety was his willingness to always carry his silver-mounted pocket pistol and a stiletto when he left the house. Although advised to avoid areas away from the manor house after dark, James Scully was unafraid.

On Sunday night, April 24, 1842, a note was nailed to the front door of Kilfeacle House which showed a coffin with a body lying in it and the legend "This is the last notice." No one inside heard a sound to alert them to danger. Around 10:00 P.M., as he was preparing to retire for the night, James Scully was shot while he was standing by a closet window near his parlor in Kilfeacle House. Five shots came through the window, one striking him "in the left cheek, which carried away three of his back teeth, injured his tongue, broke his left jaw, and lodged in his right cheek." Two bullets just missed his head, and the other two struck the wall just in front of where he was standing. The slug in Scully's cheek was successfully removed, and his recovery began. His escape from the assassination attempt was considered providential. A hundred pounds was offered as a reward for information leading to the arrest of the culprits, but it was never claimed.[23] Some arrests were made, but no one was ever brought to trial. Typically, there was a general silence from potential informers, members of the lowest class. After all, it was a case involving class against class.

In October a convicted pig thief in the Limerick gaol claimed to be one of "four armed men who attacked Mr. Scully's house," and he "named the man who fired the shot" that wounded Scully. But nothing came of this confession.[24]

In the meantime James Scully was feeling so fit by August that he embarked on a three-month's tour of the Continent. He returned to an Ireland of long winter nights in November and got to his Kilfeacle manor house late on the afternoon of Saturday, November 26. His youngest brother, William, who had just turned twenty-one years of age, was among the family and servants present to welcome James home. In the excitement and turbulence of the arrival, mention was made of ducks on the pond located about one-quarter of a mile below the house. Impulsively, James decided to go hunting; so he and his brother William quickly got their shotguns in order to take advantage of the fleeting daylight. A few minutes before 5:00 P.M. they departed for the pond. Shortly after five o'clock in the cold twilight, William, having stepped into the water, decided to return to the house. He traded weapons with James, giving him a double-barreled shotgun, as it was loaded properly for the ducks in the area; then he quickly returned to the warmth of the house. William had been out of Kilfeacle House some twenty minutes. When the usually punctual James did not return for the regular six o'clock dinner, William became uneasy and sent servants to the pond to search for his brother. They soon returned without any news.[25]

Messengers were then dispatched to the police barracks, which were four miles away at Bansha, and to the home of Jeremiah Scully, a cousin, in Golden. Jeremiah came hurriedly, and subinspector Gannon and a party of police began the search by lantern light, concentrating their effort in the large pasture adjacent to the pond. Between eleven and twelve o'clock the dead body of James Scully was found, about seventy-five yards from where William had left him. He had been shot twice in the back, and he was lying face up, "his nose broken, and face frightfully disfigured." A mallard and his India-rubber cape were found nearby. His rings, gold watch, silver-mounted pocket pistol, stiletto, and small tablet had not been touched; but the shotgun was missing, although a piece of the gun's stock, presumably broken over Scully's head, was recovered.[26]

There were extensive reports of James Scully's murder throughout the British Isles. The assassins had apparently been watching Scully closely, and they had been alerted when he returned to his manor house. News of this "wild justice of revenge" emphasized the "extraordinary sensation through the country" that was caused by this most-recent outrage in Tipperary. The government offered a reward of £200, which was augmented by pledges from others, including £1,000 from the Scully family, so that the total reward exceeded £2,000.[27] The official police report agreed with the general feeling that Scully was murdered because of his treatment of certain tenants:

27 21407 4265X 7032	52226 Tipperary S R Kilfeacle 26 Nov 1842	Clanwilliam
Homicide 24/11 42 Police Report	James Scully Esq murdered & robbed of a gun for having dispossessed defaulting tenants £200 public	£50 private reward

—Official report at Dublin Castle

Some news accounts attempted to point a moral from the ghastly conditions that brought about Scully's death. The London *Times* dwelt on the motives of the slayers, describing the crime as

> the desperate act of famished fathers, blinded to consequences and nerved to crime by the sad sight of homeless and shivering beings, to them dearer than life; and whose wan cheeks, as they pined and faded under the dark scowl of the monster rack-rent, pointed the silent finger of revenge. To men who have never known any other means of redress, or experienced any better or more effectual remedy for oppressions, than the bullet of midnight assassination, the temptation of opportunity was irresistible.[28]

The *Times* also pointed out that neither religious views nor class or rank could account for widespread assassination in Tipperary. Not only Roman Catholics but agents, yeomen, and laborers "appear indiscriminately the victims of this fearful organization," which seemed to exist for criminal acts.

Early in December a nearby landlord had the following posted on his property:

NOTICE

> is hereby given that any person who values his life or Property or that of his friends or Relatives will have nothing at all to do either directly or indirectly with certain holdings convenient to Mocklerskill lest he may meet his fate in a similar manner to that of Scullys

D.C^{e} in particular will	)	I am etc —
please note the foregoing	)	
and circulate it among	)	*Neddy Irontride*[29]
his friends ——	)	

Other reports from Cashel were associated with Scully's murder. One verbal nighttime threat was: "Take this as a warning. Have nothing to do with the lands of Mayfield or you will get the Death of Scully."[30]

From January 30 through February 11, 1843, sixty-two persons, including William Scully and his cousin Jeremiah Scully, were examined by the solicitor for the crown in an effort to build a case against the murderers of James Scully. Very little information that would be usable in a public trial was brought forward. The murder of Scully's herdsman on January 12 was unrelated to the death of his employer six weeks earlier, although it, too, was an act of revenge.[31] In 1851 the estranged wife of one Andrew Coffey accused him of the murder of James Scully. He was arrested in Liverpool, where he had lived for seven years, and was returned to Tipperary for a hearing. Since the only witness against him was his wife, who according to law could not be examined against him, he was released by the authorities.[32]

William Scully was twenty-one years of age when his brother was murdered. His father had died when he was only nine, and James had been the much esteemed oldest brother, almost a father image. His loss was a personal tragedy to the youngest brother. Their mother, Catherine, had grown old and blind, and she could not provide the leadership in the family that had been supplied by James. In this transition, Vincent became the head of the family, and William became the owner and landlord of small holdings.[33] But if conditions were bad in Ireland in the years before 1842, the ominous Potato Famine, which devastated the Irish landscape in the next few years, compounded the miseries for both tenants and landlords.

2

William Scully Becomes a Landlord

The assassination of James Scully deeply shocked his youngest brother, William, and undoubtedly set a pattern for William's actions in future years. William, as a member of the landed class, naturally took the landlord's side completely in any landlord-tenant conflict, and he was more convinced than ever that the proper behavior for a member of his class was to *demand* his legal rights.

In the meantime the London *Times* reported that some Irish landlords were attempting to ameliorate the conditions of their tenants, but for many landlords it was too far out of character to respond to the changing needs of their tenants. The *Times* sought to uphold British tradition and to keep extreme actions on the landlord's part from undermining the structure on which their very survival was based. Irish newspaper comment was closely watched and reprinted when appropriate. An item from Derry described the pretentious and haughty rejoinder from an Irish landlord which was considered all too typical of that class:

> We understand that a certain landed proprietor of a neighboring county, after having received a memorial from his tenantry, praying for a reduction of rents, and being besought to give a reply, came forth and tore the petition to tatters before their faces, telling them that was the only answer he was prepared to give, and they deserved to get, save that if he made any change at all, it would be to enlarge, rather than reduce, the price of land upon his estates![1]

James Scully, in his will, which was dated May 10, 1842, only two weeks after he had first been shot, named his younger brothers William and Rodolph to serve as executors of his estate. They were also to receive certain lands in the event of his death. When James was murdered, most of his vast estate went to Vincent, his next younger brother. William, according to a will signed by his father just before his death in 1830, was to receive land, implements, and livestock. But the property would descend to him only after his mother's death, which occurred in 1843. On March 1, 1843, several months before her death, a family agreement provided that William would receive certain lands from his brothers in "event they have no sons." Since he had reached the age of twenty-one, William received the lands of Ballinaclough, Knockroe, Springmount, Folcherstown, Donaskeigh, and Ballinliney, as provided in his father's will and verified in this deed of family compromise.[2] Catherine Scully's last will and testament, dated May 19, 1843, and witnessed shortly before her death, did not mention land, nor did it include William. That will, which was primarily for her daughters' benefit, disposed of her jewelry, household furniture, and similar personal properties.[3]

Thus, in the year of his mother's death, 1843, William got a sizable bequest, which in American money at that time would have been more than $100,000. There was a £1,000 cash settlement and land valued at £21,000. The rental income from this land was fixed at £1,300 annually, slightly larger than Denys's annual rentals in 1808. By William's own assessment this land was "exhausted, over let, cottiered and ill managed," and it was not certain that the annual income could be paid by the tenants.[4] As was the case in much of Ireland, there were too many people on too little land. Individual holdings were small, and the pressure to overuse the soil in order to gain a return without replacing the soil's fertility was ever present. Scully's use of the word "cottiered," a synonym for "rack-rented," was widely employed in Ireland to describe burdensome and extremely high rentals. Such an unreasonably high rent was based on the strong demand for rental land and was likely to be very near the full annual value of the products of the land.

Because he felt that his new estate was ill managed, William took personal control of these lands during the next seven years. He later wrote that he also farmed the worst portions, probably meaning that he directed the work of others on these estates and did not rent them out. Later, landlord Scully would be criticized in his handling of his Irish lands because of his willingness to deal directly with his tenants or to take over management roles, which, in the feeling of the times, were better left to a landlord's bailiff or steward.

However, William Scully's activities during these seven years provided him with a greater understanding of the landlord's task than had all his years of observing his older brother James. The primary change on these lands was to increase the number of livestock at the expense of crop acreage. Also, the advantage of draining the cold, wet Irish land so as to increase grass or crop production became readily apparent. Although most Irish agricultural techniques were remarkably inefficient, the illiterate Irish peasants were masters of the simple spade. They could dig ditches and keep them open for proper drainage. According to later accounts, Scully became fully acquainted with drainage techniques for swampy or boggy land.

Scully's remembrances made no mention of the severity of the potato famine, perhaps the greatest natural catastrophe to hit Ireland in all time. Between 1780 and 1845 the population of Ireland rose from five to almost nine million. Persistent population pressure from these new millions fell on the land and on agriculture almost alone, as a small Irish industry had stagnated after 1800. Irish tenant farms, often to the consternation of landlords, were divided and subdivided in order to support the new generations. Pressure on land caused rents to rise, but the smaller individual holdings meant a decrease in the total farm output.

By the mid 1840s the average Irish tenant farmed about five acres. His standard of living was lower than that of any other class in western Europe and was even lower than that of his ancestors of a hundred years earlier. Many a family lived in a windowless mud hut, which had a dirt floor and a hole in a thatched roof to serve as a chimney. The hut's interior was dirty, smoky from turf fires, and was host to a variety of vermin. Pigs and chickens, precious because sale of them helped to pay the rent, shared these hovels and added to the unsanitary conditions. In spite of heavy reliance on the potato as a food source and in spite of a lack of pure water, sewers, cheap clothing, or preventive medicine, the Irish population burgeoned. Irish boys married at sixteen, girls at fourteen, and they produced large families. Mortality rates were extremely high, but somehow many children managed to survive.

Potato production, even from one acre of poor soil, could support a large family. Ten to twelve pounds of potatoes, usually boiled and seasoned lightly with salt, constituted the average daily consumption of each Irish peasant. The almost exclusive reliance on a single food subjected the peasant and his family to the dangers of famine. Crops had failed in earlier years because of disease or bad weather. Malnutrition, accompanied by such diseases as scurvy, had brought many deaths to the Emerald Isle. Although previous famines had been of short duration, some Irish peasants had faced the threat and had emigrated to America before the years of the Great Hunger. Only fifty years earlier, Thomas Malthus had predicted that inevitably the population would grow far faster than the food supply. Ireland in the late 1840s was a walking example of Malthus's warning.

Outsiders described the Irish peasants as shiftless, happy-go-lucky, addicted to liquor, and emotional. They were stereotyped as childlike and generally good-natured but subject to violent rages. The degrading life they lived brought laziness and inefficiency, which might be corrected, so some said, by careful guidance and discipline. The poverty-stricken Irish peasant was also ingratiatingly courteous.

Such were the conditions in 1845, when a potato fungus destroyed almost the entire Irish potato crop; and 1846 was equally destructive to potato production. Snow-covered ground during much of the winter of 1846-47 hampered foraging for native plants so as to provide a meager diet. So demoralized were Irish peasants by 1847 that few potatoes were produced, although the fungus did not reappear. Even the seed potatoes had been eaten. Optimistically, heavy plantings were made in 1848, but the blight returned. The year 1849 was not as bad, and a marginal crop was harvested. Two more years were needed before potato production was back up to levels before the famine.

Almost one-third of all Irishmen either died from the famine or fled to Britain or North America. Deaths came so rapidly that many bodies could not be buried immediately. The dramatic and decisive effect of the Great Famine on Irish personality heightened Irish hatred for all things British. Irish nationalists believed that the British government sought to solve the Irish problem by inaction, thus permitting starvation to reduce the Irish population. To them it was a deliberate act. It made no difference that the British government sent much relief or that groups and individuals contributed heavily to alleviate Irish famine problems. Even Irish landlords who gave one-way tickets to tenants to help them emigrate were looked upon with scorn—naturally, for landlords held their positions and their land under British laws.

William Scully made no personal mention of the effect of the famine on his own life. As a member of the landed class he did not lack for food during the late forties, nor was he ever on the verge of starvation during his entire life. His primary criticism of the British government during this period was on an entirely different matter—the repeal of the Corn Laws in 1846. The Corn Laws had been passed with strong backing from landed interests early in the century to provide a protective tariff on grains and other foodstuffs, so that agriculture in the British Isles could be assured of higher prices. By 1846, advocates of free trade gained control in Parliament, and protective laws for British agriculture were repealed. British industrialists came into power, and they sought to make Britain the workshop of the world. Parliament thus deliberately chose to make Britain dependent on imports of food and fiber from overseas, which would permit other nations to buy the products of Britain's mines and factories. No longer would Irish grain and animals move into the British market behind a protective wall. After 1846 these agricultural products were in competition with grains and animals from the newer farming lands across the seas, especially in America. Depressed market conditions for Irish agriculture added to the trials of famine.

It was William Scully's fate to commence his trials as a landlord during these disastrous years. He committed all of his energies to his new estates. He discussed landlord practices with other landlords, and he read all that he could find on successful agricultural pursuits. His labor and his capital were invested in his lands, which were said to exceed a thousand acres, placing him among the upper 10 percent of all Irish landowners. By January 1, 1850, he figured that his cattle, sheep, horses, implements, and growing crops had a market value of about £1,400. Moreover, during these exceedingly hard times he had saved about £2,000, represented either by cash in hand or money loaned. Proudly he proclaimed in an account written in 1879, "I owed nothing!"[5]

In William Scully's youth, land was the key to Irish life, and the competition for it was brutal. The great landlords were Protestants whose ancestors had gotten their lands from the English king. They, in turn, rented much of their land to great tenants, who sublet to others, who sublet smaller tracts to the lower classes. At the bottom of the social heap were the one-acre cottiers, reaping a bare margin of existence from their small holdings. Generally, landlords owned only the land—tenants developed the needed improvements. Most Catholics held land for the life of a particular tenant or were tenants at will and were subject to eviction for failure to pay rents on time or for any other offense. Rental agreements might cover many years or a single year. Insecurity of tenure was a basic reason for small investment in improvements. The Scully family, for instance, did not renew their lease on the Kilfeacle lands, because they believed that the rents that they paid as lesser gentry had been increased too much. In later years the manor house where William Scully was born was demolished. The scramble for land and the subdividing of holdings were products of the rapidly growing population. After the Great Famine, further subdividing diminished as emigration to the city or to a foreign land absorbed a growing population. William Scully quickly adapted to the idea of finding opportunities in a new land, even though he was not from the classes that were most likely to use this expedient.

Clues regarding Scully's relations with his tenants during his first few years as a landlord are scant. Much was made later of an account published in a popular history, *New Ireland,* which stated that in 1849 Scully "was tried at Clonmel assizes for the shooting of two fine young men, named Bergin, sons of a tenant whom he was evicting at Ballinaclough; but he was acquitted on this charge."[6] No doubt these were the sons of either William or Thomas Bergin, who together had a lease of "two lives" on sixty-seven of the Ballinaclough acres at an annual rent of £107 7s.[7] The Clonmel spring assizes for the eighth day of March, 1849, recorded five criminal charges:

Number		
175	William Scully & Jeremiah Scully shooting at Will Bergin to do grievous bodily harm	True Bill
176	William & Jeremiah Scully shooting at John Bergin	do
176½	Jeremiah Scully & Thomas Twoomey shooting at Will Bergin . . .	do
180	William Bergin (the younger) Stephen Bergin shooting at Jeremiah Scully with intent to do him some grievous bodily harm	[Not informed]
181	William Bergin and a person unknown same	True Bill[8]

The outcome of these charges cannot be found in the official record. Apparently all the persons listed were acquitted, and it is interesting to note that William Bergin appeared in four of the five charges either as the aggrieved party or as the aggressor, as did Jeremiah Scully, William's cousin.

Trouble with tenants may have caused Scully to look elsewhere for an investment sanctuary for his capital and his agricultural knowledge. His confidence in the new English policy was shaken because of Parliament's repeal of the Corn Laws. Unrest in Ireland also led to an unsuccessful revolt in 1848 that touched county Tipperary. Not only economics but the birth of a son by an unmarried young woman on August 30, 1849, in Dublin gave Scully a desire for a change. The boy was named John Scully, and Scully later assumed the responsibility for his education and eventually provided John with an important position in his landed estates.

Attracting Scully to other climes were the reports that he received from the United States. Good virgin land existed in an almost limitless supply at low prices. Moreover, he was particularly interested in the invitations to "men with capital" that were coming from America. Typically, persons who responded to the lure of the land in the United States were members of the lower middle class or even the lowest class, if they could get the means to travel. They had no opportunity to possess land and the political and economic power that went with it in the "old country." Most persons like William Scully, as members of the upper class, had an enviable position already, and there was little inclination to risk that in a new place where old associations might be meaningless. Thus, the attractions of the United States, however great, did not generally draw persons from the privileged classes in other lands. William Scully was an exception to the type of person who came to American shores in the mid-nineteenth century: he had economic power and social position and the promise of continued privilege in his homeland.

When most persons with William Scully's background compared both the propelling forces that urged removal from old haunts with the advantages of remaining and the attractions coming from the new land with the risks and hardships involved, the decision was easy—stay put. Scully, on the other hand, wanted to see for himself. He had a vision that his wealth, limited as it was, could be greatly augmented. The secret, he believed, was in selecting low-cost land of high quality, and he believed that his experience and knowledge would enable him to do just that.

Later, Scully was to explain it this way:

> Just at the time when I was seeking fields for my work I was attracted to the United States. I was a farmer. I had worked in the field on our Irish estates, and had learned the economics of agriculture at schools. So I was more than a farmer. I was a scientific farmer.
>
> The people of the United States were at that time very eager to induce persons with brains and money to come and assist them in the development of this wonderful country. I was interested. I knew I had the money, and I was pretty sure I had the brains. A man must be sure of this to be successful. Reputable newspapers were printing invitations to Europeans to come. . . . I came upon the solicitation of the United States government.[9]

Drawn by stories of soil quality and cheap land in the United States, William Scully booked passage on the Cunard Line's steam packet *Britannia,* which landed him in Philadelphia early in 1850. He went west on the newly constructed Pennsylvania Railroad to the end of the line, near Altoona, Pennsylvania. There he purchased a horse and saddle, and he had a blacksmith make him a small iron spade for use in sampling the prairie soils.[10] The spade could easily be carried in a saddlebag, and if one made a handle from an available tree, it would be convenient and serviceable. Similar spades were employed by other land investors at the same time.

Thus equipped, Scully set out on a systematic investigation of farming lands in Pennsylvania, Ohio, Indiana, and Illinois. The object of his summer-long search was to find fertile, productive, low-priced land that could be obtained in large quantities. Wherever he went he dug small holes in the lands that he was considering. He wanted to know how deep the top soil was, and he believed that he could recognize soil quality through visual inspection.

After much deliberation, William Scully made his initial selection of United States land in central Illinois. He had come into the region from the northeast to Middletown, had proceeded on the Edwards Trail to Springfield, and had traveled through the center of Logan County. Tradition says that Scully and his horse almost drowned in coming through part of the swampy region in northwestern Logan County where he decided to buy land.[11] Scully attracted little attention in these exploratory investigations. He "appeared quietly in Lincoln, where he engaged board." Using Lincoln, the Logan County seat as his base of operations, he gave "no one an intimation of his intentions. Every day he was on horseback, scouring the prairies."[12]

Scully later declared: "I not only selected the richest soil I could find, but I secured land in a straight line between the cities of St. Louis and Chicago. I knew there would be a railroad between those cities, and that rich prairie land close to the railroad would be valuable." He also said, "I secured my land very cheap."[13] He could have added that his first land purchases were in one large block.

Illinois became a state in 1818, and Logan County was organized in 1839; yet, eleven years later this land was still available for purchase from the federal government. The high land and forest land in the county had been claimed many years before and was occupied. As a matter of fact, the cold, wet prairie to which Scully was drawn was the only large block of government land remaining in the immediate area.

Scully's land purchases were made at the Springfield Federal Land Office on October 11 and 17, 1850. Twenty-seven entries were made on each of these days, for 4,320 acres on the eleventh and for 4,200 acres on the seventeenth.[14] The cash price for this government land was $1.25 per acre, which would have amounted to $10,650, but Scully paid for it with military land warrants, land scrip that could be obtained for less than ninety cents an acre, thus greatly decreasing his personal cash outlay. These initial purchases probably cost Scully less than eight thousand dollars, which, with his expenses, would have exhausted most of the savings that he had accumulated over the previous seven years.

A congressional act of February 11, 1847, authorized land bounties to veterans who had a year of more of military service in the Mexican War. Conceivably a veteran could obtain his grant at any district land office, but few cared to take their land directly. Although there was discussion in Congress about making these warrants unassignable, they could be transferred to a new owner before being used to buy "offered" land, and the properties acquired by Scully fit that description.[15]

Elated by his success in acquiring a princely domain that was larger than the amount of land possessed by any of his brothers, William Scully quickly retraced his route back to Dublin and to Tipperary, where he informed his family of the opportunities that awaited their capital in the United States. He sought to enlist his brothers in his enterprise. While they were interested, they did not want to leave the life they were leading in Ireland. Thomas, William's next older brother, was most receptive to William's plea for support, but his immediate interest was in riding to the hounds, something he would miss out on if he were to take a journey to the New World. Thomas agreed to lend his younger brother the immense sum of £10,000 for use in purchasing land.[16] William got together all the money he could raise without mortgaging any of his Irish estates and then made plans to return quickly to the bountiful lands of Illinois, where his agricultural experience could be put to work.

3

Scully and His New American Lands

The year 1851 was a momentous one for William Scully. He added more than twenty-one thousand acres to his Illinois holdings, which would be the basis for his tremendous wealth. That was the year of his first marriage, and in November he reached his thirtieth birthday.

By March, Scully was back in Illinois after his hurried trip to Ireland to obtain more money. Much of the time since the previous October had been spent in travel. With funds lent by his brother Thomas, Scully made his purchases again at the Springfield land office, first on March 29, 1851, for twenty-seven tracts, containing 4,260.89 acres. Three and one-half weeks later, on April 22, he entered his largest purchase, 9,815.92 acres in sixty-seven parcels. Later that week he filed on another nine tracts totaling 1,410.59 acres. He made no entries in May, but on June 3 he purchased eight tracts for 1,280 acres and on the twenty-eighth, twenty-one pieces of land amounting to 3,247.54 acres.

By that time Scully was becoming well known in Springfield as a buyer of huge quantities of land, and he could never again move unobtrusively in the Springfield area. Large purchasers of government land, such as Scully, were received cordially in the federal land offices. The receiver and the register, the primary officials in a federal land office, were paid in part by commissions on land sales which were based on the total acres disposed of. One buyer of thousands of acres could be handled much more expeditiously than many buyers of small parcels of land. Thus, purchasers like Scully were given special treatment, and the prevailing attitude was that they deserved that consideration.[1]

Although two additional tracts were purchased in Scully's name that year, one on July fifth for 169.40 acres and one on September nineteenth for 160 acres, in all likelihood Scully had returned to Ireland by that time. Patents of ownership, which were issued by the General Land Office in Washington, for the lands entered the previous year were beginning to come in by that summer. Except for several small purchases, which were canceled due to previous entry or other error, Scully received patents on these early entries within a year.

Table 3.1 shows the lands purchased by William Scully at the United States Land Office in Springfield in 1850 and 1851, the first full year of his American activities.

TABLE 3.1

SCULLY'S PURCHASES OF ILLINOIS LAND, 1850 AND 1851

Date	Number of Entries	Acreage
October 11, 1850	27	4,320.00
October 17, 1850	27	4,200.00
March 29, 1851	27	4,260.89
April 22, 1851	67	9,815.92
April 26, 1851	9	1,410.59
June 3, 1851	8	1,280.00
June 28, 1851	21	3,247.54
July 5, 1851	1	169.40
September 19, 1851	1	160.00
Military Warrant Purchases	188	28,864.34
Cash Purchases	12	664.04
Grand Totals	200	29,528.38

SOURCE: "Abstracts, M.B.L.W. Locations, Act of 1847, Illinois," vol. 36, National Archives.

Such large purchases at the land office gained recognition for the twenty-nine-year-old Irish land buyer. Community sentiment in those days applauded landed investments in public lands as promotion of the area's economic growth. However, because Scully was frequently seen in the Springfield Land Office, there was criticism of him on the grounds that he was engaged in "sharp practice" in entering land. For instance, it was alleged that he listened carefully to men who were making inquiries about land and whether it was available. If they did not buy immediately but went off to examine it more carefully, they might return to find Scully's name as purchaser on "certain tracts of land upon which they had set their heart."[2] They were sure that Scully made these entries because of their remarks, on the assumption that they were the best lands available, not because he might have seen the lands himself.

Scully again enhanced his cash resources by using Mexican War military land warrants for the purchases acquired in 1851. Ten of his quarter sections had excess acreage of 95.11 acres, for which he paid the cash rate of $1.25 per acre. In addition, twelve tracts of various sizes, adding up to little more than a section, were paid for with cash under the act of March 24, 1820.[3] Still, he could have gotten all of this land with cash because of the amount of money lent to him by Thomas, but he retained much of it in order to protect his investment. Moreover, there were expenses connected with land acquisition, especially when one did not wish to resell immediately or when one had traveled as far as he had. There are family stories that for a short time Scully maintained an office in Springfield to acquire land scrip, but no operation of that sort can be found in the contemporary record. However, Springfield newspapers regularly contained advertisements for the purchase and sale of soldiers' land claims, and Scully probably patronized these businesses.[4]

During this first full year of American land acquisition, all of Scully's land was concentrated in Logan County, most of it in one large block in the northwestern part of the county, with a small amount in adjacent Tazewell County. Because his holdings were so large, they appeared attractive to other land seekers, but the area selected by Scully had been overlooked for a generation. Other users of military land warrants at Springfield during the same period who acquired large quantities of public land were John D. Gillett, Roswell P. Abel, Robert B. and other Lathams, James C. Conkling, Christian P. Anshutz, John L. and Thomas Officer, Abijah Taylor, Jr., and John Williams.[5] Possibly, Williams, who was a Springfield merchant, was the best known to Scully. Williams served as Scully's banker and agent on many occasions, and when Scully began purchasing land west of the Mississippi River in later years, he made a large purchase from Williams.

William Scully probably returned to Dublin in midyear 1851 to take part in his marriage to Margaret Mary Sweetman, whose home was on Fitzwilliam Square North, Dublin. On May 6, 1851, while Scully was still in Illinois, a contract known as a Settlement of the Tuter Marriage was drawn up in Dublin. Where considerable wealth was involved, such antenuptial settlements for disposition of properties were typical. William's Dublin address in this contract was 5, Great Denmark Street. The bride, a daughter of Michael Sweetman, a wealthy Dublin brewer, brought a dowry of £5,000 into the marriage. The marriage agreement stipulated that the bride was to receive two annual cash payments totaling £500. Scully's Irish estates—at Donaskeigh, Ballinliney, Ballinaclough, Knockroe, Springmount, and Folcherstown were listed, but not his American lands. Cosigners of this contract, in addition to the bride, her father, and the groom, were Michael James Sweetman and Thomas Scully.[6]

Contemporary evidence fails to disclose when the marriage took place or when the young couple moved to Illinois to make their home in the middle of the newly purchased land. But they must have been there on March 24, 1852, when Scully bought fifty-five tracts, involving two large blocks of land in southern Grundy County, at the Chicago Land Office. Payment for the land was again made with the use of military land warrants. Margaret Mary's dowry could have accounted, in part, for this final purchase of 8,792.36 acres of public land in Illinois.[7] Scully's holdings in Illinois then totaled 38,320.74 acres, an area equivalent to one and two-thirds of a congressional township, or just under the total acreage of sixty sections.

Contemporary observers were generally critical of the land that Scully bought, and they cited many of the land's shortcomings. Typically, settlers with bad experiences on bottom lands misjudged these low-lying lands in Illinois. The notes of the official federal survey, which were based on observations made in September, 1823, show that most of the land acquired by Scully was wet, swampy, level or rolling bottom land, although the soil was first rate.[8] A traveler in 1831 pointed out that the region was very attractive during seasons of little rainfall, but he cautioned that "the emigrant may mistake a dry season, and fancy he has a rich, level and dry farm in prospect, but the next spring will undeceive him."[9] A county map legend in 1844 described the area as "barren cold bottom land; unfit for cultivation."[10] In spite of this foreboding evidence, however, Scully could see that this land would be worth far more than what he had paid for it. In the next few years, as a matter of fact, he bought land from private sellers in Mason, Will, Cook, and Grundy counties and another 792 acres in Logan County.[11] Initially, he seems to have envisioned that he would develop his holdings under his own personal guidance.

Near the head of a branch of Prairie Creek, on approximately one thousand acres in the center of his Logan County lands, Scully set to work to make his land fruitful. Here, on what he called South Farm and North Farm, he did much to transform the area. South Farm, which had 543 acres, soon had two excellent wells of stone and brick, fences with posts charred to avoid rot and set two feet deep, a barn forty feet long by thirty feet wide, a young orchard of apple and peach trees, and a fifteen-acre locust grove of about five thousand trees. North Farm, slightly smaller at 535 acres, was on a good living branch of Prairie Creek. This farm had about eighty acres in timber. A hundred acres of level, beautiful land was quickly broken, and an excellent dwelling, costing between twelve and fourteen hundred dollars, was erected. When it had been painted white, it was referred to as the Big White House or the Scully White House. The Chicago and Mississippi Railroad had been completed by the time Scully built his house, and the airline distance from this house to the Lincoln depot was slightly more than ten miles. Though Sugar Creek and Kickapoo Creek provided obstacles for travel from Scully's farms to Lincoln, the county seat, local tradition holds that Scully's home became a stopping place for newly arrived Irish immigrants. Some families may have come with him, and many others gravitated to his place to work for their countryman.[12] Such reports intimate that William Scully's reputation among newly arrived Irish immigrants in the 1850s was high and that many were acquainted with him from the old country.

Crops were planted on both South and North farms. Because of the swampy character of much of Scully's property, where the local name of Delavan Swamp was quickly converted to Scully Swamp, most of the soil could not be worked until June or later each year. Only under rare conditions could a good crop of corn be raised. Thus, raising of crops was restricted to knolls or high points of the land. Apparently Scully had anticipated difficulty in cropland agriculture, since he quickly expanded the livestock on his land with a large flock of sheep and a few head of cattle. Many of the sheep were imported from the British Isles, and the size of his operation can be seen in an August 18, 1853, advertisement in the *Illinois Journal* (Springfield): "800 sheep for sale. About 500 wethers, from 2 to 4 years old, and about 300 ewes, from half to full blood Merino. Between Sugar and Prairie Creeks, Logan County, Illinois."

A major hazard for sheep-raising, in addition to the wet countryside, was the existence of a large number of prairie wolves in the area. Scully found it necessary to build tight corrals and to pen his sheep at night.[13] Local tradition does not recall that much effort was made to drain the area in the first few years. One important technique of draining—the use of a mole ditcher—was not employed in Illinois until 1854 or later.[14]

By 1853 conditions looked very promising for Scully, so he made two decisions that suggest that he had made up his mind to locate permanently in Logan County, Illinois, United States of America. On June 7, 1853, he went to the Logan County circuit court to file a declaration of intention to become an American citizen.[15] Such a declaration was an initial step in the naturalization process, which took five years, but he failed to gain citizenship because of illness in his family.

His other 1853 decision was the preparation of an elaborate and detailed "Map of the Estate of Wm. Scully Esq situate in Logan & Tazewell Cos., State of Illinois," which was completed by a draftsman in October, 1853. Overall, Scully's map, at a scale of two inches to a mile, covered six governmental survey townships, with small segments of four others. Along the right-hand margin of the map was the town of Lincoln, while San Jose was in the upper left-hand corner. The route of the Chicago and Mississippi River Railroad was featured on it, as well as a projected direct air-line route from Lincoln to Delavan, with an added notation "R.R. w'd probably go 1/2 mile west of this."[16]

Since Scully had his map copied from official survey plats and since he later added features from his own tissue-paper plats, he noted any discrepancy turned up by his own surveys. These numerous brief notations in Scully's handwriting provide more information about his character than do other sources from this period. He was precise in his notes and was ready to correct official surveying errors. Although his words were few, they contained much information about the land. For example, Scully entered a note along Kickapoo Creek: "Difference in Plats on site of creek"; and along Sugar Creek he wrote that the blazing was "examined with J. Switzer & Hendrickson & chained & the creek correctly altered here. Taking from Wisemans stake May 22 & 23 /54 Wm. S." Along Prairie Creek he showed: "Creek accord'g to plat" and "Creek accord'g to my survey."

Scully carefully scrutinized fences and corners along his external boundaries. Sample entries in his writing were: "Fence [within] 5 feet of line," "fence on line," "fence nearly on line," "old ploughed line about on line," and finally, "I think Pooles fence runs this way," and he drew a fence line angling seven or more degrees out of plumb. Most often, corners were noted as "post & stone," although a variety of markers were employed, such as "forked post with one nail" and "fallen stake & large rock 4.5 yds. N. of Branch." For another corner he wrote, "I put walnut stake here 1760 yds. E from Town'p line." On another small tract he wrote, "I put two small stakes at points A & B . . . and blazed the trees as far as the sleugh only, with J. Switzer & H. Hendrickson—May 23d /54 Wm. S."

Natural landmarks were identified on Scully's map, with locations such as hills, groves, waterways, ponds, sloughs (which he spelled "sleughs"), and lone trees. Extra information was provided in comments such as "groves paced and examined with compass," "lone tree about exact," "high hill," "low hill exact," "excellent pond," "hill surveyed," "grove exact." In the extreme corner of his land was a small "sand hill." Outside his land boundaries, Scully noted: "Range of Hills," "cherry grove," "sand hills. pure sand."

Scully's estimation of the quality of his land was shown by many of the map entries. The single word "flat" appears on many tracts. "Low rich rosin weed land," "flat rosin weed land—very rich," and "rich flattish land with sleughs and hills" provide examples of Scully's idea of soil quality. He also labeled some land near a range of hills as "bottom and swampy land," and another quarter section was shown as "wet & very flat [with] low hill." Near a "good deep branch" he wrote, "A very choice 1/4"; and along Sugar Creek in one place he noted, "Bottom black sandy loam." Near a "bad sleugh" north of Kickapoo Creek he recorded, "The worst 1/4 I have."

Man-made features were also reflected on Scully's map. Houses were shown, accompanied by the word "exact" or by a statement verifying the location that had been made by someone who knew the area. Bridges, crossings, a rocky ford, and roads were marked. Also shown just outside the Scully lands were a saw mill, a forge, and a school.

Three daughters were born to William Scully and his wife, Margaret Mary. They were Mary Gertrude, Julia, and Kathleen. The eldest may have been born in Illinois, but information about these children is so scanty that little survived in later years.[17] It is known, however, that the damp, humid, unhealthy conditions around the newly built Scully White House so undermined the health of Mrs. Scully that late in 1854 she left Illinois, never to return. Apparently she had contracted malaria, which in certain forms was prevalent in the swampy regions of central Illinois. William went with her to Ireland and to France, where she sought to regain her health. Plans were made to put Scully's affairs in Illinois in such an order that the landowner's presence would not be required. Agents could pay the taxes and maintain the value of the properties.

In 1855 Scully returned to Illinois to work out procedures for the use of his land, including its possible sale. It is conceivable that Scully signed up his first tenant, James Hickey, a recent Irish immigrant, in 1854.[18] In August, 1855, Scully advertised a dispersal sale: "1200 sheep; about 600 ewes, 200 wethers & 400 lambs, young healthy, & in good condition. Also 30 head of cattle, consisting of cows & calves, yearlings & a fine bull."[19] Knowledge that he was departing sparked the interests of numerous buyers, and offers arrived even before he sold off his livestock.

A Mr. Straut (or Straught) apparently made a firm offer for all of Scully's unimproved Logan County land in April, 1855. In a fashion that was typical of his character, Scully carefully set down calculations about the advisability of selling by the piece or *in toto.* First he considered the proposition for twenty-eight thousand acres, excluding the improved land of his North and South farms. He had paid close attention to the increasing value of land, and he believed that his land was now worth five to five and one-half dollars per acre; so he wrote down $20,000 as an essential down payment. Additional income from such a sale was figured on the basis of 6, 8, or 10 percent interest on the remaining balance, with payments of the principal to be made at the end of two, four, and six years. Such calculations produced a minimum of $208,000 to a maximum of $240,000, which was many times his cost only three or four years earlier. Realistically, Scully wrote that he might be expecting prices from fifty cents to a dollar an acre too high.

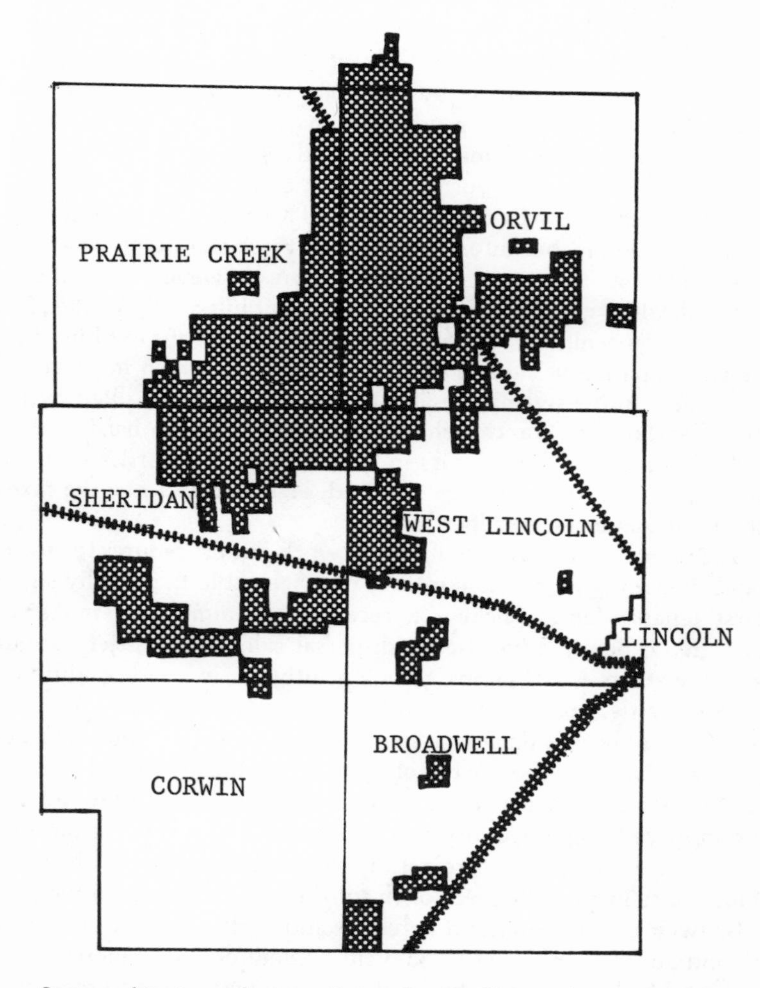

Six townships in northwestern Logan County, Illinois, showing concentration of lands owned by William Scully in 1900. His original purchases here were made in 1850 and 1851. The slight projection north of these six townships shows Scully land in Tazewell County.

At the same time, the Irish landowner projected returns from retailing his land—175 quarter sections—which he believed could be sold from 1855 to 1861. Terms would be granted to prospective purchasers, and prices would be increased on new land sales as the years passed. Even with increasing costs, due to taxes, private expenses, and the trouble in handling at least 175 sales, Scully believed that retailing of his land would produce a 10 percent greater income.[20] Although there is no way of knowing why Scully opted to sell his land piecemeal, the chances of wholesaling it may have fallen through, or perhaps he preferred to handle things himself. In fact, he was already selling to small purchasers by the time of Straut's offer to buy the entire unimproved acreage.

Between April 11 and December 20, 1855, Scully or his appointed representative drew up sales contracts for more than three thousand acres. Some land was completely paid for upon signing the contract, but usually there was a down payment with provision for payment of the balance within three, five, or even seven years. Cash down payments varied from one-fifth of the total price to one dollar or one dollar and fifty cents per acre. The cost per acre was usually larger for small tracts. Interest on unpaid balances was due semiannually, usually at the rate of 8 percent, with 10 percent obligated on overdue notes. Buyers also agreed to provide capital improvements on the land, amounting to several hundred dollars, during the first two years of the contract, a requirement that would discourage large-scale speculators. Each buyer also became responsible for all taxes on the land.

Sales were better in 1856, with two contracts drawn up in January and twenty-five signed during the last six months of the year. The land area sold was up 57 percent, and the average price per acre increased even more. Selling opportunities looked good for 1857, but only five contracts were written—two in January and one each in March, May, and July, when Scully abruptly terminated his effort to sell his Illinois land. The Panic of 1857 brought a price collapse on all forms of business, and the years from 1857 through 1862 "brought uncertainty and retrenchment of original plans" to other investors in Illinois land.[21] Table 3.2 summarizes Scully's land sales from 1855 to 1857.

Except for one or two cases, the names on these fifty-two land contracts indicate that Scully was selling land to other men who were of Irish stock, came from American families of long standing, or had roots in England or Scotland. The lands placed under sales contracts were from the perimeter of the largest block of Scully land or from outlying tracts. The 1853 map was updated with changes in ownership through January 17, 1857.

Scully had put all of his land up for sale, even North and South farms.

TABLE 3.2

CONTRACTS FOR SALES OF SCULLY'S LANDS, 1855–57

Year	No. of Contracts (Acreage)	Price per Acre	Average Price	Number of Sales Not Completed (Acreage) Price	Number of Sales Completed (Acreage) Price
1855	20 (3,129.57)	$ 4.50–11.00	$ 8.88	7 (851.67) $ 7,049.65	13 (2,277.90) $20,673.90
1856	27 (4,929.88)	10.00–20.00	13.48	15 (3,701.64) $51,702.00	12 (1,228.24) $14,790.64
1857	5 (400.21)	12.00–20.00	15.08	4 (360.21) $ 5,244.00	1 (40.00) $ 800.00
Totals	52 (8,459.66)	$ 4.50–20.00	$11.85	26 (4,913.52) $63,995.65	26 (3,546.14) $36,264.54

SOURCE: "Tax book of William Scully—Sales of Land," Office of the Scully Estates; Gates, "Frontier Landlords," p. 179; Official Records, Office of the Logan County Recorder, Lincoln, Ill.[22]

Sales on one-half of the contracts, involving less than half of the land, were actually completed, and Scully later repurchased some of that land at a much higher price. The inability of buyers to comply with the sales contracts was due to a variety of conditions so typical of the frontier. Only in the first year of sales, when the price per acre was smaller and down payments were bigger, did a majority of the buyers complete their contracts. Perhaps Scully was more interested in selling land at that time, as there is no other positive relationship between the time of sale, the acreage involved, and completion of the contract.

Scully's recourse when a buyer was unable to complete his contract was to treat the down payment as rent and to collect interest on the outstanding principal. Court records show that he frequently extended purchase contracts for two or more years. For instance, the sale of 80 acres to J. W. McBride on September 5, 1855, required an $80 down payment, a payment of $350 principal three years later, and a final payment of $370 on September 5, 1860. The agreement "in witness whereof the said parties of these presents here have here unto interchangably set their hands" called for 8 percent interest on the balance, to be paid annually, for payment of all taxes from 1855, and for improvements worth $150 to be put on the land within two years. McBride received a short extension before his death, and his heirs made a new agreement on March 11, 1861. On December 1, 1863, they forfeited some of their rights, but on October 19, 1870, the deed of transfer was finally recorded, based on the amended contract of 1861.[23]

In a similar manner, William Willis signed a purchase agreement with Scully on September 1, 1856, for 160 acres at $10 per acre. The final payment was to be made five years later. The contract was extended for two years at higher interest, and Willis had paid Scully in full by June 11, 1863.[24] William Hickey, on the other hand, was a party, along with William Scully, to numerous official records in the Logan County Recorder's Office. After signing a promissory note for $100 and a judgment note in 1857, Hickey apparently contracted to buy a tract of 40 acres from Scully in 1859. The land contract was voided by court action five years later.[25] Hickey was also a Scully tenant during this period, paying share rent with a portion of the corn crop.[26]

Two early-day Logan County lawsuits, which were typical of those involving William Scully, may account in part for later stories that "Scully settled numerous tenants upon his lands by agreeing to convey tracts to them if they would make certain prescribed improvements. The lands were thus brought under cultivation and some were conveyed to tenants, but many of the contracts were never fulfilled and the rights reverted to Scully."[27] In a case involving a land sale, the defendants claimed that they had paid $1,800.00 on a $1,993.25 note, whereas Scully's agent maintained that nothing had been paid. The court held that the defendants owed $2,508.71, counting interest due to time the case was initiated. Added interest and court costs brought the total judgment to slightly more than $3,550. Two years later the money had not been paid, and the land reverted to Scully.[28]

The defendant in another Logan County lawsuit was tenant John McMullen, who had rented land in 1855 just north of North Farm. Believing that McMullen was about "to depart from this state with the intentions of having his effects removed from this State to the injury of . . . William Scully," Scully's attorney brought an attachment suit against McMullen for $135.50, the rent for that year. After Scully won this case, McMullen also had to pay Scully's cost of $81.80 and other court costs.[29]

During the 1850s William Scully showed an unusual combination of flexibility, restraint, and adaptation in the development of his American estates. He carefully picked his land for the quality of its soil, for its sizable acreage, and for its potential in regional economic growth. At the same time, he did not put all of his extensive capital into low-cost land, but he retained a large part of it in cash for use in protecting and developing his holdings. At first he tried to supervise his estates personally; but conditions changed when his wife became ill, and he therefore made plans to sell everything. However, the terms of sale and the worsening of economic conditions caused another alteration in plans, which moved in the direction of agent-managed tenant operation of his estates. After the 1850s, changes would be made in Scully's enterprises, but never again did he show such abrupt innovation in his plans in order to adapt to new situations.

After Scully's return to Ireland in 1855, he initiated a schedule of visiting his American properties every other year. Because of such prolonged absences from Illinois, it was necessary to appoint representatives to act for him. As early as mid 1852 he had given a power of attorney to John Switzer to act for him in the sale of a right of way to a railroad. There is no available evidence to substantiate a direct appointment for John Williams, leading merchant and financier in Springfield, but he was given custody of Scully's tax book and certain assets. Williams, who acted as Scully's banker and honored his drafts whenever they were presented, was specified as early as 1856 in Scully's sales contracts as the man to whom payments were to be made.[30] Between 1856 and 1858 Scully's account with Williams had sums on deposit ranging from $3,000 to $8,000.[31]

Samuel C. Parks, a leading attorney in Lincoln, handled many of the details involved with the Logan County properties. Parks also served as a local agent for Williams, collecting debts owed to the Springfield merchant. Scully had valuable property to protect, and he particularly detested having to pay delinquent taxes, so Parks made "a distinct arrangement" with the county collector "that no cost is to be made on Mr. Scully's lands & that before advertising or making out his [delinquent] list to advertise he is to notify me when he has leisure to make out the Receipts for Mr. Scully."[32] Never in these early years did Scully have a full-time agent to oversee his American holdings. William McGalliard, a law partner of Parks's, also served as Scully's agent and was his primary contact in Lincoln after 1863.[33] When Scully consulted lawyers other than Parks or McGalliard, it was about specific questions or on certain lawsuits. For instance, Judge Stephen T. Logan, a well-known Springfield attorney, advised Scully concerning the accuracy of some of his land patents, and the Chicago firm of Shumway, Waite and Towne handled one lawsuit in Logan County.[34]

After his return to Ireland, Scully maintained his residence at Ballinaclough, near Golden in county Tipperary. He returned to Illinois in 1857, 1859, 1861, and 1863; and a pattern of renting his land began to emerge. Most published materials on Scully indicate that he began to rent his land in Illinois to tenant farmers during the Civil War, but the evidence shown in this chapter points to small-scale leasing almost as soon as he built his home on North Farm. Most of these rental agreements were an individual matter between the landlord or his agent and the tenant and did not possess the universal characteristic of the one-year cash-rent leases that were typical of Scully lands in later years. Most of these leases did specify cash rents, but on some the landlord received a share of the crop.[35]

Unoccupied land was frequently settled by squatters or was exploited by neighboring landholders, so Scully developed plans to attract tenants with five-year leases that had graduated rentals. For instance, the tenant's cost for the first year would be an amount equal to taxes, with nothing for the landlord. The second year it might be twenty-five cents per acre, increasing within a few years to a rate of one dollar or more per acre. Usually, capital improvements—such as breaking the sod or erecting fences—were required. Other improvements were left to the tenant.

An example of a more formal early lease was one drawn up for five tenants in 1864.[36] The 560 acres that were involved had been returned to Scully because of a voided sales contract, and the five-year lease was written at an annual rate of one dollar per acre. Each tenant was assigned a specific area, and he agreed "to cultivate [the] land in a good and husband-like [manner] and to do and suffer no waste to be done thereon during this lease—to take good care of all houses, fences or other improvements there may be upon said lands." Also, in the lease, the landlord denied responsibility for crop damage due to poor fences, nor would he accept liability for building or repairing fences.

During this early period, letters passed between William Scully and John Williams, and some can be found more than a century later. Scully's letters show an intense interest in agriculture and a pessimism about the income coming from the Illinois land. Comments in these letters were brief but businesslike. For example, in 1860 he reported that "there are little or no manufacturers in Ireland, and the agricultural & pastoral interests have passed this a terrible hard & dry summer & long wet & hard winter & spring, the worst in 13 years, & prospects [are] not too good." Later that year he was more optimistic when he wrote that agricultural activities had improved and "prospects in trade in England [are] encouraging."[37]

Reports coming to Scully from Illinois were more favorable, and shortly after the outbreak of the Civil War, Scully asked Williams to transfer to him $7,000, payable in gold. The problem of transferring large sums of money across international borders was a recurring one for Scully in later years, but he showed his concern when he wrote of "indications in Chicago that all currency under par or under 90 cents will before long be rejected—and a strict specie or specie basis prevail."[38] Throughout this period, Scully persistently maintained that he was losing money on his Illinois land, and he treated the $7,000 transfer in 1861 as capital, not income. In 1863 Scully questioned the price that Williams charged for preparing a certified check, which was one-fourth of one percent higher than the charge on a similar check in Chicago. Income taxes, imposed in the United States for the first time during the Civil War, also became the subject of several letters of 1863. Scully first wrote to Williams that his income for 1862 was about $800 less than his taxes and other expenses. Just before he departed from New York City on the ship *Persia,* he remembered receiving about $500 for his own corn crop of that year, and he figured that his income was $300 less than his expenditures, which he said were $4,820. He relied on Williams to prepare his income-tax report, "as all my receipts pass through your hands but not all my expenses—and you have the best opportunity of furnishing a return."[39] Neither in that year nor in other Civil War years did Scully pay any income tax, although his income from leasing in Illinois was beginning to grow. Scully's letter to John Williams dated June 22, 1865, closed: "With felicitations for the successful termination of the war and the better prospects before us, & with best wishes for your welfare I remain."[40]

During the period from 1851 to 1865 the procedure employed on the Scully estates in Illinois was primarily a holding operation. Some income was derived from sales of land during a brief period from 1855 to 1857, after Scully had abandoned plans for personally developing his real estate. Because of taxes and other continuing expenses, he maintained that he had to rely on income from his Irish lands to keep things going in Illinois. After 1855 he transferred his home and family back to Ireland, and his personal interest in his lands seemed to go there. He had to pay off his debt to his brother Thomas, and he had growing expenses because of his daughters and his wife's extended illness.

However, the manner in which William Scully developed his Irish estates in the 1850s and 1860s made the Scully name notorious in Ireland and well known in England. By 1870 William Scully had become a symbol for oppressive landlordism. Irish peasants had turned against him, and he almost lost his life. Land legislation enacted by Parliament in 1870 was said to have been in response to actions that Scully had taken on his Irish estates.

wswswswswswswswswswswswswswswswswswswswswswswswswswswswswswswswswswswswswswswswswswswswswswswswswswswswsws

4

Bloodshed at Ballycohey

wswswswswswswswswswswswswswswswswswswswswswswswswswswswswswswswswswswswswswswswswswswswswswswswswswswsws

William Scully turned his back on Ireland and his position on his landed estates when he established his residence in Illinois and entered his declaration of intention to become an American citizen. But his difficulties in productively converting his virgin lands in the New World and the illness of his wife changed all that. In 1854 he took his family back to Ballinaclough House, located near Golden in county Tipperary. In an effort to improve his wife's health, he took his family to the sunny southland of France. Personal family reasons, therefore, altered Scully's conception of his landlord role in Ireland and his relation to his lands in Illinois.

Part of Scully's motive for abandoning Ireland to go to America in the early 1850s was the repeal of the Corn Laws, which caused a great reduction in Irish farm income on sale of grains on the English market. In reconsidering his objections to existing market conditions, Scully sought to implement innovations on his Irish estates that would respond to new economic situations. In essence, these alterations were designed to make a large reduction in the numbers of tenants, a move that was bitterly resisted in Ireland in the 1850s and 1860s. But in Scully's view, the only salvation for Irish agriculture was to improve the fertility of the soil and to produce animals and animal products for the nearby growing urban markets; and some landowners agreed with him. Such a move would drastically curtail the acreage in cultivated crops in favor of pasture and would produce an enormous increase in livestock. Since efficient handling of livestock required fewer tenants than older patterns of cropland agriculture, the tenant who might be expelled from his lease or the one whose relatives would be forced to move could be expected to resist.

Few nineteenth-century landlords would have consulted with their tenants over issues of this sort, and it was William Scully's style to view his future plans for his Irish estates as a private matter that was of no consequence to the tenants whose lives would be altered by his decision. But the nature of the Irish lease, which usually was made for years at a time, did not permit a sudden move of this nature to take place; instead, if a landlord were to make changes, they would occur over the course of many years. Yet, Scully was an impatient man, with a strong feeling that his prescription for his ailing lands was the correct one.

In Ireland there was an "undeclared but most real Agrarian War, the war of Tenant *v.* Landlord, of the Irish People *v.* the English Garrison, of Ireland *v.* England"; and Scully, though an Irishman with an English mother, strongly sided with his class in society—the landlord.[1] Half-hearted parliamentary efforts to solve the Irish problem produced discussion of a land bill in 1852, which was abandoned only after almost violent opposition from landlords. Palmerston's Land Act of 1860 provided that landlord- tenant relations be "governed by contract," which sounded good for a tenant who had faced arbitrary action by his landlord, but it was in fact a strengthening of existing powers of landlords. Thus, the landlord's right was virtually absolute over his land; he could set any condition and impose almost any rent. At cross purposes with this new legislation was the underground fight for Irish nationalism, which used land rights as a last-ditch struggle for survival of Irish civilization.

A basic disagreement in Irish land issues had been the use of English law, which held that tenants "had no proprietary rights in the land," whereas Celtic traditions "assumed they possessed rights of that kind." While the Irish felt that an "occupancy-right of tenants-at-will" was disposable property which could be sold to strangers, the English would not accept that view. Nevertheless, without English sanction, a tenant right did continue, and its price varied inversely to the rent. Sometimes even an English landlord would pay an outgoing tenant in order to get a "right of quiet possession." Such tenant right to property was bequeathed in wills, often amounting to more than the value of the land to the landlord. Rents, at the same time, were usually set according to criteria that were customary, rather than financial, and hereditary landowners generally opposed increasing the rents. After the potato famine of the 1840s there were strong efforts to curb tenant right to the land, even where tenants owned the drains, fencing, and buildings. British opinion, conditioned by English tenant-landlord roles, failed to comprehend why there was so much agrarian outrage in Ireland. It did not seem compatible with the usually high moral standards of

Irish peasants. But to the impoverished Irish peasants the "title-deeds of nearly all the landlords who opposed the tenants' demands were traceable to confiscation." Even in the 1860s the Irish, in talking of "very recent" confiscations, were referring to happenings of two and three hundred years earlier, and they were using such remembrances to justify agrarian violence. To backers of Irish nationalism, assassination was a weapon to use in gaining their goal. Killing a landlord was murder only in a technical sense.[2]

In this period of increasing hostility between Irish landlords and their tenants, the brothers and sisters of William Scully were pursuing their own careers. Vincent became head of the family when James was murdered in 1842. Like James, he was trained in the law and in 1833 was admitted to the bar in Ireland, where he soon built up a big practice. He became a member of the queen's council in the 1840s and was awarded the distinction of a "silk gown" in 1849.[3] In 1841 Vincent was married to Susan (or Susannah), daughter of John and Sarah Grogan of Mayvore, county Westmeath. His residences were at 2 Merrion Square, in Dublin, and at Mantlehill, Castle Park, near Golden.[4]

In a letter to the London *Times* in 1847, Vincent maintained that he had good relations with his tenants, in spite of "the present state of the country." In response to an important question of that period, he did agree "that the houses of the landed gentry should be provided with sufficient arms for protection in cases of emergency." As a landlord of twelve years' experience, Vincent professed to have no knowledge of any conspiracy among his "tenants against the payment of rent."[5] His letter to the editor carefully stated his position in such a responsible way that it is not surprising that Vincent was elected to Parliament from county Cork, sitting from 1852 to 1857, and reelected to serve from 1859 to 1865. But he had financial problems during this period. In 1857, when the Tipperary Bank was bankrupt, Vincent, as one of the bank's owners, became engulfed "in a sea of litigation," which forced him to dispose of much of his vast estate, which amounted to 3,116 acres in counties Tipperary and Cork.[6]

The next brother, Rodolph Henry, had property in the Kilfeacle area, but he maintained his residence in England. He was married in 1850 to Mary, daughter of John Graham Lough, a noted London sculptor. Her mother was the daughter of the Reverend Henry North, who had served as chaplain to the duke of Kent. Three years earlier, in 1847, the oldest sister in the Scully family, Catherine Julia Mary, had been married to a Spanish duke, whose name was Prince Antonio Publicola Santa Croce, duke de Carchiano, duke de Sante Gemini, count de La torre, grandee of Spain of the first class.[7]

"Dear brother Thomas," as William referred to him, was next. He willingly lent a fortune to William to help him buy Illinois land in 1851.

Thomas's estates at that time were extensive, and his Irish holdings were slightly larger than those owned by William. Like his oldest brother, James, he never married. Thomas, a skillful horseman who was vitally interested in sports activities, had a freak accident at the age of forty and subsequently died. Newspaper accounts from Cork say that it was a riding accident from which he did not recover because of "the fatigue he underwent during the [Parliamentary] contest for this county."[8] Thomas bequeathed his Irish lands to William, thus more than doubling William's estates in the Emerald Isle, but ownership of these new lands brought him serious trouble. Both of William's younger sisters, Mary Anne and Juliana, became nuns in the Society of the Sacred Heart and spent most of their lives in religious service in France and Belgium.[9]

After William Scully returned to Ireland in the mid 1850s, he sought to rebuild his life as an Irish landlord. Since he signed or notarized deeds of sale of some Illinois land in this period, a partial record of his movements can be reconstructed. Deeds for 1855 were notarized in Chicago, whereas in 1856 both Scully and his wife appeared before the county Tipperary commissioner of the high court of chancery to provide valid signatures. In 1860 the sale deeds were signed either in Golden or in Dublin.[10]

In the ten years before January 1, 1860, Scully's wealth had increased more than fourfold. Showing an increase of £4,000 from his personal evaluation of 1850 was his "own old Irish Estates," which he thought were worth £25,000. The Irish holdings left to him by Thomas were given a value of £30,000. The "balance of about 35,000 acres in Illinois [was] worth this date about 8 dollars per acre," or a total of £56,000. Investments "in cattle, sheep, horses, farm implem'ts and crops grown (chiefly hay)" on his Irish lands had more than doubled to £3,000. His "cash on hands" was about £1,000, and he had no debts. As he reconstructed it in later years, of his estimated wealth in 1860—£115,000—about half was in his holdings in Illinois.[11]

In 1861 Scully made plans for his biennial visit to Illinois. While he was gone, his wife, Margaret Mary, went to southern France to seek a climate that would be more conducive to her frail health than any that could be found in cloudy, cool Ireland. On May 20, 1861, when she was endeavoring to return to Ireland, Margaret Mary Scully died in Avignon. When Scully, in far-off Illinois, heard of his wife's death, he hurried back to Ireland, leaving many decisions to be made through correspondence. His young daughters needed his care, and he turned his attention even more to his Irish estates. In black-edged letters to John Williams, Scully remarked about the recent death of his wife. His blue stationery for the next two years was black-edged, a traditional sign of mourning.[12] Scully's visits to Illinois in the rest of the 1860s were rare, possibly limited to 1863 and 1866. His land-buying and -selling days in Illinois were virtually over; his name does not appear on the Grantor-Grantee Index in Logan County for the years 1865 through 1869.

Perhaps Margaret Mary had been a moderating influence on landlord Scully's handling of his Irish lands. In any event, it was after her death that William Scully was a party to two serious conflicts with tenants—conflicts that might have been avoided if someone whom he respected had questioned his actions. Reports of Scully's land activities in these later years were often highly colored by "hostile" sources, modified only in part by explanations presented by the landlord himself or by his agents. Never after 1865 did he enjoy favorable press coverage, and some of the people who were close to him presented evidence that was damaging to landlord Scully's reputation. The first of these serious landlord-tenant conflicts was at Gurtnagap in 1865, where William Scully came close to killing a tenant's wife. The second was the Ballycohey affray in 1868, which resulted in two deaths; Scully's was almost a third.

Gurtnagap (or Gortnagap), in county Kilkenny, was some eleven miles west of the town of Kilkenny, near Tullaroan. Located on the Munster River in the heart of the Slieveardagh Hills, the Gurtnagap farm, which was in rougher terrain than most of the land owned by Scully, was part of the estate bequeathed to William by Thomas. One hundred and ten acres at Gurtnagap were farmed by five related Teahan families under a lease drawn up by William's father in 1820. Edmond Teahan, the last life on the lease, died on August 8, 1864, and Scully then sought to gain control of the land in order to implement his plan for larger production and more efficient use of his land. In October, 1864, making no mention of his plans, Scully accepted rent for the Teahan-farmed land at Gurtnagap, without granting a lease. Rent paid in this manner would be for one year at a time, and ejectment papers would have to be served by formal writ. In the spring assizes of 1865 Scully sought ejectment of all of the Teahan families, but the court decided against him.[13]

Shortly after Scully paid the costs of this lawsuit he appeared at Gurtnagap and took up residence about May 9 in a room in the house of John Caldbeck, his bailiff or caretaker. Ejectment writs were ordered from Dublin, and they arrived in late May. Such notices, to be legal, had to be delivered in the home of each tenant to be ejected, in the presence of the tenant or a member of the family who was over the age of twelve. Typically, ejectment writs were delivered by the landowner's bailiff and were perhaps witnessed by local police. Earlier Scully's agent at Gurtnagap had not been able to find tenants at home to serve writs, partly because they knew that the landlord would not compensate any of the tenants for their improvements on the land. They believed that Scully would clear the land with his "Crowbar Brigade," thus destroying their possessions of a lifetime. Moreover, they wanted to retain their homes. So the Teahans at Gurtnagap organized to prevent service of the detested ejectment writs by keeping their gates and their house doors locked and by refusing to answer summons to open up. The law provided little protection for tenants, but if they could hold out past the end of May, the earliest that Scully could bring them into court on ejectment proceedings was during the following spring assizes. Thus, they would have possession of the land for another summer and winter.

During this May-time war of nerves, Scully, with several bailiffs and two policemen, was frequently seen around the Gurtnagap land. The police were equipped with rifles and bayonets, and Scully carried a revolver and dagger. John Caldbeck, Scully's bailiff, later said in court that he "could not say that the tenants are very fond of Mr. Scully—(laughter)—he used occasionally to go into the houses and look around them; never heard him say, 'By your leave,' when he went in; he used to go in, I think, like a wolf, look round him, and then stalk out again—(laughter)—Mr. Scully used to be armed; the arms were in a belt, inside his coat; always had a couple of policemen with us to look respectable before the people—(laughter)—."[14]

Finally, on May 29, 1865, ejectment notices were served on all of the Teahan families. The most difficult service was on the home of Patrick Teahan, which was found locked early in the morning and again in late afternoon. Finally, about nine in the evening, Scully, with his agents Caldbeck and Patrick Quinn, made his way into the Patrick Teahan yard, leaving two subconstables to stand guard out on the public road. Around ten o'clock the front door opened, and Bridget Teahan, Patrick's wife, went down a pathway to the front. Caldbeck, Scully, and Quinn then moved forward to serve the writ. Mrs. Teahan recognized them and spread the alarm. She also returned quickly to the door to pull Caldbeck out of the house and to grab Scully by his cape, ripping its hook out of the eye. In the confusion and turmoil, Caldbeck served the writ, and Scully hit Mrs. Teahan over the head with his walking stick.

Blood streaming from a three-inch gash in her crown, the hysterical Mrs. Teahan ran to the police barracks, which were about one-quarter mile down the road, yelling, "Murder, murder!" Scully and his party followed, as did many of the members of various Teahan families. Out of this brief encounter came three lawsuits, two especially that were damaging to Scully's career and reputation. Two of the cases were civil suits: one was a county court case of *Teahan* v. *Scully*, asking for £1,000 damages; the second was an ejectment suit of *Scully* v. *Teahan*, heard in the city court. The third was a far more serious criminal case, *The Queen* v. *William Scully, Esq.*, which was tried before the county crown court.

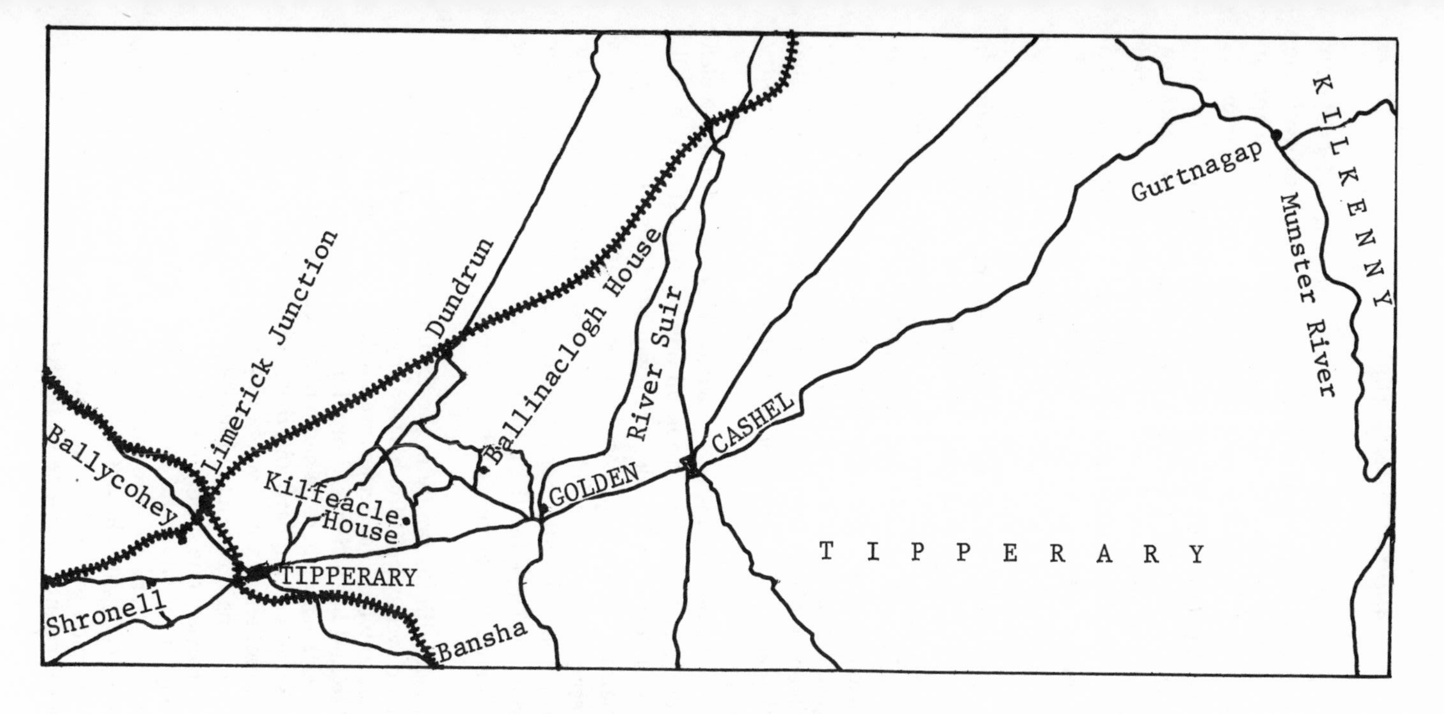

Portions of counties Tipperary and Kilkenny in southern Ireland. Ballycohey, at the extreme left, is thirty-one miles from Gurtnagap, at the extreme right. The towns of Cashel, Golden, and Tipperary, as well as the locations of Kilfeacle House and Ballinaclough House, appear on this adaptation from a modern Irish Ordinance Survey map.

Mrs. Teahan's condition was described as critical from May 29 through early June. Scully stayed at Gurtnagap after the confrontation with the Teahans. He was arrested on June 3 and was taken several miles to appear before justice of the peace John Waring, Pottlesrath House, "who committed him to gaol, bail being denied," because doctors still maintained that Mrs. Teahan's life was in danger. The London *Times* called this an agrarian outrage, committed by a landlord. By June 11 Mrs. Teahan had recovered somewhat, and Scully was released. The complaint of Mrs. Bridget Teahan against Scully was presented at the petty sessions in Kilmanagh on July 14, and Scully was remanded for trial at the August sessions of the Kilkenny court.[15]

A Kilkenny newspaper headline during this period proclaimed: "Ejectments—Shocking Assault of a Woman—Arrest of Wm. Scully, Esq." Later publicity on these cases was presented in a comprehensive but less sensational manner.[16] Altogether, the three cases in which William Scully was a party were reported extensively in more than two hundred column inches in the early August issues of the Kilkenny *Journal.* Similar coverage appeared in the Kilkenny *Moderator,* with lesser comment by the Cashel *Gazette and Weekly Advertiser,* the Clonmel *Chronicle,* and the *Tipperary Express & Advertiser.*

The first lawsuit to be heard was the case of *Teahan* v. *Scully,* which sought damages of £1,000. This trial took all day on August 1 in the Kilkenny county record court, with Judge O'Hagan presiding. The three attorneys for the plaintiffs opened the case by stating that Scully exhibited a considerable bitterness toward the Teahans because of the outcome of the previous ejectment case. The attorneys tried to build the feeling that the landlord was seeking revenge for the affront he had suffered. In part, the newspaper transcript of their presentation reported:

> Mr. Scully lived with his family in the County Tipperary, where he moved in the society befitting his wealthy position and enjoyed all the luxuries to which gentlemen are used; but shortly after the ejectment case had been disposed of what did he do? He went to the lands of Gurtnagap where he had had previously a room fitted up in the house of his own herd and there took up his abode for a considerable time, leaving his own family residence where he habitually lived, leaving the society of his family and friends, leaving the home which every one should hold dear. . . . Mr. Scully so far forgot the relation in which he stood to the Teahans—he so far forgot the duty which he owed to the position of a gentleman who moved in an exalted rank—he so forgot the solemn obligations imposed on every man who was a landlord and accustomed to deal with his poorer countrymen—he so far forgot everything of the

> kind as to undertake in his own person the duties of a common bailiff.[17]

The point was also made that "it was only the mercy of God that Mr. Scully was not . . . standing in the dock arraigned before the Court for having killed and slain Bridget Teahan, of Gurtnagap."

In endeavoring to explain Scully's motives in the ejectment case which was basic to all that came later, solicitor Serjeant Armstrong asserted that William Scully

> was a highly respectable but queer gentleman. . . . [meaning] that he was a determined man in the assertion of his own rights. If he thought, as he did think, that his tenants had put their heads together and entered into a combination to retain possession of the land in contradistinction to his will—then Mr. Scully became determined in the assertion of his own rights; and it was this that induced him to go and see that ejectments were served on the Teahans.[18]

Some effort was made to show that the Teahans had conspired to place Scully in an awkward position and that much of the response to Mrs. Teahan's scalp injury was mere acting.

Many questions in this first trial dealt with the weapon used by Scully in striking Mrs. Teahan. A doctor described the wound as having been made by a sharp instrument, and a Teahan attorney held that it must have been made by a saber. Mrs. Teahan remembered it as a short, dark, heavy object, while Scully vowed it was a walking stick, which he had had for three or four days before the May twenty-ninth assault. His witnesses described it as an oak stick, no bigger around than the middle finger, about two and one-half to three feet long. Scully said he threw the stick away several days after the beating, and he could not find it. The police were criticized for having neglected to seize the weapon, which was not produced at the trial.

The plaintiff's summation for the jury, whose members were peers of the defendant, emphasized the real sufferings of the peasant's wife, condemning Scully as "a disgrace to the class which owned him—the landlords; and concluded by calling for a verdict which would show to the world that a special jury of the county of Kilkenny were anxious to protect the poor, shield their rights, and discountenance oppression."[19] News commentary reported that there was "loud and protracted applause" following these statements.

The judge's charge to the jury was lengthy in "clearly and forcibly laying down the law on the essential points of the case and most eloquently expatiating on the impropriety of Mr. Scully, a landlord, going to his tenants' farms and acting in the capacity of bailiff." After a short absence, the gentlemen of the jury found in favor of the Teahans. But instead of the £1,000 asked, they were awarded £80 and assessed court costs, which were set at 6 pence. The Kilkenny *Journal*, in a mocking editorial, declared: "This assault case . . . was a remarkable case, and we venture to say, for many reasons, it will be long remembered—*by the people*. . . . Bravo, gentlemen, you have immortalized yourselves: what landlord now will hesitate to indulge in the luxury of fracturing the skulls of his tenants' wives when he can do it at the cheap rate of £80 a piece!"[20]

The criminal case of *The Queen* v. *William Scully, Esq.*, was tried before Justice O'Brien in the Kilkenny county crown court on August 2, 1865. After pleading "not guilty" to charges of "violent assault on Mrs. Bridget Teahan," Scully was not questioned at all. Examined and cross- examined were Bridget Teahan; Doctors Ryan and Delany, who had attended Mrs. Teahan; Constable Anglin; and John Caldbeck, who was Scully's bailiff. The questioning about the alleged assault took much the same line as the suit for damages of the previous day, except that Mrs. Teahan also claimed that she had had a miscarriage on July sixth due to the injury that she had suffered. Dr. Ryan corroborated this statement, at least in part, by saying, "On the 6th of July Mrs. Teahan had a miscarriage; at least I was told so; she had hemorrhage at the time; that would have arisen from miscarriage; or from other causes."

Reading the account of this trial more than a century later, one has no way of knowing why Scully was not called to testify. Perhaps his counsellors felt that the hotheaded, impulsive landlord would not make a good witness on his own behalf. One of his attorneys, in a long address to the jury, described the difficulties that the landlord was having with his tenants at Gurtnagap. He emphasized that Scully was "on legal and proper business, while Mrs. Teahan was illegally resisting the process of her Majesty's courts of law" and that Scully had given the blow in self-defense. Nevertheless, the jury felt otherwise, finding Scully guilty; and he "was then given into the custody of the gaoler."[21]

Sentence for the conviction of William Scully in the assault of Bridget Teahan was passed down by the judge on the next day, August 3. Scully's attorney, in a long address to the court, presented an affidavit from the prisoner which provided details leading to the assault. He stressed the fact that Scully "deeply deplored his rashness on the occasion," that the blow that struck Mrs. Teahan was not premeditated, that it was by a "stick and a stick only," and that he was sorry that they could not produce the stick in court, which if "shown to the jury their verdict would have been perhaps quite different." The counsellor further stated that "for the last twelve months, the unfortunate prisoner had suffered from ill health, he was even then in such a very delicate state of health that any imprisonment would be attended with most injurious, if not extremely dangerous consequences." Justice O'Brien, at length, stated the various points of law and described the contradictory evidence in the case. He pointed out to Scully that "the circumstances mentioned in your affidavit as regards the state of your health, are, . . . more properly for the consideration of the executive than for the consideration of the judge." Finally, the judge presented this sentence:

> In every wish not to pronounce a punishment at all excessive, in every wish that if, in measuring out punishment, I should err in giving less than I ought to give, than in giving more than I should, I do not think the ends of justice would be satisfied by pronouncing on you a less punishment than that to which I am about to sentence you; and that is that you be imprisoned for *twelve calendar months* from the first day of this assizes.[22]

Perhaps Scully's plea for mercy may have been of influence, for he maintained later that he did not serve the sentence imposed at Kilkenny. Whether his health was a factor or whether the case was appealed cannot be determined from the available materials. His descendants had a feeling that probation must have been granted and that he must have paid a fine in lieu of imprisonment.[23]

On the same day that sentence was imposed by the queen's court, the Kilkenny city court, under Judge O'Hagan, heard the ejectment case of *Scully* v. *Teahan.* Because twelve gentlemen jurors were unavailable, ten served with consent of counsel. This short case hinged on a section of the landlord and tenant act and on how it was affected by the rent paid the prior October. After hearing the evidence, the judge ruled against Teahan's attorney and "directed the jury to find a verdict for Mr. Scully." Scully had finally won a case, but execution of the ejectment was delayed for three weeks, upon the request of the attorney for the defendants.[24] The first three days of August, 1865, had indeed been busy ones for Scully. Would he learn that extreme behavior had certain limitations?

After lengthy coverage of the trials, the Kilkenny *Moderator* editorialized:

> Indeed, in matters both criminal and civil, the lawyers were on this occasion most largely indebted for briefs to one individual, Mr William Scully, whose litigious propensities brought him before the courts in various positions, leading to his paying rather dearly, on the whole, for the indulgences of those propensities. If

> he gained one civil suit he lost another; and at the criminal side of the court he was the sole sufferer. If damages in which the jury mulcted him in the civil case which he lost, were lighter than some folk seem to have expected, there can be no mistake that his sentence, under the criminal prosecution, was severe enough.[25]

Comment from the *Freeman* appeared a few days later in the Kilkenny *Journal,* which blamed "bad and unsound" laws that encouraged landlords to take advantage of tenants. It further stated that "Mr. Scully and his class are not yet sufficiently enlightened to see that their own and the public interests suffer from that policy, and we must attribute to him and to his sympathisers the blame that attaches to themselves personally and no more, and allow our legislators to bear the remainder." The *Freeman* further depicted Scully as having used "an amount of over zeal which excited an irritable temper, and led to the consequences under which he is now suffering." Also, it said that "the aristocrat who ambitioned possession of the poor man's house and of his own land never thought of offering to buy from the tenant that which was his." The final editorial remarks were: "We hope the result will cause him and others to conclude that it would be better policy to compensate a tenant for his improvements than to crush him into the grave in an effort to satisfy the craving for the property of another, which an iniquitous law has implanted in the heart of nearly every Southern Landlord."[26] It was all too true that Scully had paid a bill of £80 damages and more than £300 for total expenses connected with the trials—to the time of his sentencing—and that was far more than the value of the tenants' possessions at Gurtnagap. A final comment in the *Journal* appeared in late August, before Fenianism and cases involving the free Ireland movement took over the news. It sarcastically reported that Scully's behavior suggested that he was totally "right as a landlord—i.e., to do what he liked with his own, and treat tenants as vermin."[27]

In evicting tenants, Scully was following closely a pattern of other Irish landlords, although he may have used the law more fully in these actions, as he impatiently sought to speed ejectment. Between 1858 and 1870 close to fifteen thousand tenant families were evicted in Ireland. Fenian activity was increasing throughout the period, and Scully was criticized by these nationalists. There is no way of knowing what his brothers were saying to him, although one of them later could find little support for Scully's widespread evictions. It is known, however, that William Scully was admonished by his parish priest for his full use of all his legal authority against his tenants. Perhaps these first warnings against his arbitrary and sudden evictions were made privately, but in time, William Scully was censured in a public remonstrance. Scully's response was to place his small daughters in his wagonnette and go to the established Protestant church.[28] A century later an Irish columnist, in commenting on Scully's career, wrote, "Catholics made some of the worst landlords in Ireland"; and another author maintained that Scully was "burdened with the added unpopularity of being an apostate Catholic."[29]

Scully's leaving the Catholic church after the incident at Gurtnagap and before that at Ballycohey brought an extra dimension into the tragedy at Ballycohey. Most accounts of the Ballycohey affray are based on A. M. Sullivan's chapter in his popular history *New Ireland.* Sullivan described the reception of the Catholic congregation after a Sunday mass, when they heard that Scully had gone off to the Anglican church in this way: "They took off their hats and gave 'three cheers'; delighted that 'Billy,' as he was called, was no longer one of themselves."[30] In addition, William Scully's departure from the Catholic church estranged him from his brothers and sisters, a separation that became permanent and isolated him from the kind of support and loyalty that can come from families.[31]

The shooting and bloodshed at Ballycohey was also of significance in prompting Parliament to pass a measure in 1870 to provide a partial solution to the Irish land problems.[32] If Gurtnagap made William Scully notorious in Kilkenny and Tipperary, his actions on his Ballycohey estate attracted more widespread attention from the entire British Isles. Because of what happened at Ballycohey, Scully was increasingly identified as "an infamous landowner in the County Tipperary," as "the terror of the unfortunate tenantry," and eventually as "evictor in excelsis."[33] Scully was not wrong in determining what should be the eventual use of his lands, but his impatience and headstrong attitude in how these plans were to be carried out, in addition to his resolution in personally implementing them, almost cost him his life.

The fertile, productive lands of Ballycohey (or Ballycohy) in county Tipperary were granted to James Scully, grandfather of William, by the Right Honorable Joseph Lord Baron Milton, on a lease of three lives on October 30, 1782.[34] The Scullys, as intermediate landlords, paid £800 per year for the 321 acre Ballycohey townland, and in turn they leased it to farmer-tenants for annual per acre rentals of £3 5s. In 1839 Carbery Scully, a cousin of William's, took over supervision of this lease, and with the death of the third person named on the lease in 1847, it went back to Milton's heir, a Lord Portarlington, who reduced the rents because of the potato famine.[35]

Ballycohey was located on highlands three miles west of Tipperary town and within two miles of the important railroad connection of Limerick Junction. The property exchanged hands in a sale by the encumbered estates court in 1855, when it was purchased by justice of the peace Charles Grey Grey for the earl of Derby. Twenty-two tenant peasant families were living at Ballycohey in May, 1866, when it was sold to William Scully for a reported £12,000 to £14,000.[36] The new owner believed that this area could be supervised easily in conjunction with his other holdings in Tipperary and Kilkenny. He was building a trained and dedicated staff to oversee his estates. Darby Gorman, a young man whose father had been Scully's steward and bailiff, had taken over that job. Young Gorman, an able and attractive man, had been sent by Scully to Edinburgh University for training in veterinary and agricultural science. Gorman felt a heavy obligation to the landlord because of that generosity, because he was being paid £100 per year as steward, and because his widowed mother and his younger brothers and sisters were permitted to remain as tenants on a portion of Scully's land near Golden.[37]

About eighteen months after obtaining possession of Ballycohey in November, 1867, Scully drafted a new lease for his tenants. He was ever mindful of his vision of pasture land instead of cropland and of few tenants instead of many. The new document, which was long and complicated, sought to strengthen the landowner's rights at the expense of security for the tenants on the land they farmed. No tenant signed the new lease, chiefly because of the "clause requiring them to give up their farms at twenty-one days' notice."[38] Also, tenants were obliged to renounce claim to their improvements and even to their growing crops.

Again, in June, Scully tried to obtain tenant compliance by notifying the Ballycohey people that he would accept their rental payments at the Dobbyn's Hotel in Tipperary. Not as foolhardy as his brother James had been twenty-six years earlier, William had applied for police protection. Two subconstables were assigned as his bodyguards on a twenty-four-hours-per-day schedule. When Scully received the rent on the appointed day, he was heavily armed with two loaded revolvers, and an armed policeman was standing nearby. His intention in Tipperary was to get a signature from each tenant on a new lease or to present the tenant with a writ of ejectment. Fearing this, most tenants sent their rent with someone else, and no one signed the new lease or stayed long enough to be served with a writ.[39] Heated words were exchanged between rent payers and the landlord.

Not to be thwarted in his plans for his land, Scully's next step was to organize a party to carry out his threat and to serve writs of ejectment at the houses of the tenants at Ballycohey. Two extra police were obtained, and with his regular two-man police guard, his steward Gorman, and his herdsman Maher as the driver of his carriage, Scully appeared at Ballycohey on Tuesday, August 11, 1868. When the landlord's seven-man heavily armed party was seen approaching the area, signals were passed on from field to field, most of the houses were hastily abandoned, and an angry, noisy, unruly crowd gathered. Shouts of "Murderer," "Robber," "Have his life," and he will "never enjoy Ballycohey" were directed toward the landlord, who carried a revolver and a double-barreled, breechloading shotgun. After four or five ejectment writs had been served, the crowd swelled in number to more than a hundred and pressed and jostled the landlord's party. Maher, left with the vehicle, was beaten, and he retreated to safety. Two peasant farmers named Dwyer and Greene, with some success, sought to calm the crowd in order to prevent a breach of the peace. Finally, the pressure of the crowd was so great and the anger so evident that the constable in charge suggested to Scully that his party withdraw before it was too late. Reluctantly, he agreed, and his group returned to the carriage, where it found that Scully's traveling bag had been stolen and that Maher was hiding nearby.[40]

By the time the party got back to Tipperary, on the way to Scully's home at Ballinaclough, word had preceded them, and they were met by an excited, angry crowd of women and children. So threatening was the mob that Scully retreated to a hotel. Additional police were called out to disperse the hostile crowd and to clear the streets.[41] The weather was so rainy on Wednesday and Thursday that Scully sought no activity at Ballycohey on those days, and everyone had time for breathless anticipation. In the meantime, plans were being developed by the landlord and by his tenants that would produce a climactic struggle on Friday, August 14, 1868, the most completely reported day in William Scully's long, eventful life.[42]

Scully applied to subinspector Saville of Tipperary for the protection of additional policemen for Friday and was given four, under the command of Sergeant Daniel Cleary of the Tipperary station. Cleary would bring with him subconstables Colleton, Cahill, and Morrow. There was an understanding that Scully would receive the number of constables needed to protect his life, and their only orders were to do just that. Also, Scully would have with him subconstables Magrath and Kelly from the Dundrum constabulary, who had been assigned as his guards for "several months past, without intermission." In addition, the landlord would take his steward Gorman and herdsman Maher. Maher was notified by Scully on Thursday evening to be prepared to go some distance and to be ready early the next morning. Darby Gorman had gone to the village of Golden, so Scully gave the message to Gorman's mother. Gorman, having premonitions of what the morrow would bring, spent a restless night, ate a breakfast his mother had prepared, and, in his room before he departed, put a note saying, "I shall never return to this house alive."

At Ballycohey, some of the tenants decided that their only alternative to Scully's new lease was to resist with firearms. The house of John Dwyer (also known as O'Dwyer) was picked as a site for an ambush, because the house and adjoining barns formed a solid L-shaped group of buildings around a courtyard, with hedgerows occupying the other two sides. Fourteen men and boys—Michael O'Dwyer, Michael Hanley, Patrick and William Quinn, John and Timothy Heffernan, John Ryan, Denis and Laurence Hayes, John and Patrick Greene, John Hanrahan, Kenneth Twomey, and Michael Foley—formed the heart of the conspiracy, but virtually everyone at Ballycohey knew that something was being planned. All the firearms, some ancient and unserviceable, that could be found were gathered up. Dwyer's place was fortified and loopholed, and escape routes were planned. John Dwyer and several of the Ballycohey elders would go to Tipperary at the appointed time to establish personal alibis.

In August the sun rises early in Ireland. By seven o'clock, Scully had assembled his small party in the cool mists at Ballinaclough. Accompanied by Gorman, Maher, and his two police escorts, he went south about six miles to Bansha station of the Waterford and Limerick Railway, where they caught the train for Limerick Junction. There they were joined by constable Cleary with his three subconstables, Colleton, Cahill, and Morrow.

About nine o'clock, after leaving the railway yards at Limerick Junction, Scully's nine-man party went west along the railroad tracks to the ice house; then they turned south in the direction of Ballycohey. This unusual entrance to Ballycohey necessitated their crossing several fields before they got to Scully's land. Upon reaching Ballycohey about nine-thirty, they were immediately observed, and whistles and shouts brought "from every side the peasantry . . . trooping down in a promiscuous throng of men, women and children." Soon more than a hundred people tried to block Scully's path. Epithets of the previous Tuesday and some new ones, such as "Scully, the bastard," were shouted at the landlord, and only two or three of the legal writs to quit could be served. The constable in charge gave warning that Scully would be protected in any way necessary. The opposition to the "death notices" from the landlord caused one peasant to proclaim loudly, "We might as well be dead as alive—what can we do?" This statement should have alerted Scully that his tenants were desperate, willing to go to any lengths to oppose his mandates.

Constable Cleary, noting that the mob was increasing, went to Scully and advised him to leave. Scully asked, "Is that your opinion?" When given an affirmative reply, Scully responded with, "Well, now for the junction."

The mob was closing in, and retreat seemed to be the better part of valor. Police with fixed bayonets cleared a path. But on the way back to the railroad station was the house of John Dwyer, one of the most defiant tenants. Frustrated by most of his activity of that day, Scully announced, "We will try this one," and his party headed down the walled *boreen* leading to the house. Several guards stayed by Dwyer's outside gate to hold back the crowd.

The house and barn of John Dwyer at Ballycohey, where William Scully was gravely wounded while he was trying to evict the occupants.

From all appearances the house and the barn were empty. Scully, Gorman, Maher, and constable Morrow went into the quadrangle behind the house and made their way toward the door. Just as Scully and Gorman entered the house, the trap was sprung, and two shots rang out. A third shot came at Gorman from a porthole in the barn. Scully was the primary target of the firing, and other shots were directed at him, striking him on the head and neck. He fell to his knees, but recovering quickly, he rushed back to crouch behind the pier of a gate near the corner of the courtyard, where he began to direct his revolver fire at the windows of the house and at the porthole in the barn. Sergeant Cleary, seeing that Scully was wounded, then brought his constables into the enclosure, where rifle fire was directed at smoke coming from the porthole in the barn.

From behind the pier, Scully asked Cleary, "Will any one enter the house with me?"

Cleary responded, "I will, come on"; and they rushed the house. Just inside the door was the dying Gorman, but it was otherwise empty.

Scully said that there were people in the loft, and Cleary answered, "Very well, I will go there." As the ladder was gone, he jumped on a table to look into the loft. It was soon evident that many preparations had been made for the ambush. A feather tick in the loft had been set up for concealment and protection, and a hole in the rear thatch had been made to offer escape. None of the defenders of the house were in sight, but a half-empty bottle of brandy, a glass, some bullets, and a four-barreled pistol, with one barrel empty, were found.

Scully and Cleary then rushed the barn, saying, "There must be someone inside." Together they forced in the barricaded barn door, but they found that a back escape door was open. Near the freshly made porthole was a blunderbus that had burst. The firing had ceased; no defenders were found; and an inventory of the frightful episode was made.

Gorman was dying, having been hit at least twice in the first seconds after the house door was opened and by additional shots later. Subconstable Morrow was found dead near the porthole, with six wounds on his body, one of them in his heart. Maher was lying on the ground nearby, a bone in his leg shattered by a bullet. Subconstable Kelly had received a thigh wound, and subconstable Cahill had a slight scratch over his heart, where a bullet barely penetrated his tunic. Colletin received a shot on the back of his head which drew much blood. Scully was bleeding profusely from five wounds on his head and neck. Only Cleary and Magrath were untouched—the seventeen or eighteen shots fired by the defenders of Ballycohey had taken a heavy toll. All the parties in the ambush escaped without having been seen. Whether any had been hit by the fire of Scully's party was questionable to the investigating police, but one may have lost a hand when the blunderbus blew up. Later reports acknowledged that there had been several wounds from grapeshot. After retreating from the site of the ambush, the defenders went out to the assembled crowd and mingled with them. For a short time they would be protected, but when rewards for their capture were issued, they all got out of Tipperary as soon as possible. Several joined the British army, and others went to America or to Australia. Michael Dwyer, the "hero" of the resistance, fled to America and returned to Tipperary after twenty-five years.[43]

Years later, one of the defenders at Ballycohey recalled that "Scully was the bravest man there."[44] Nevertheless, the most popular account of the tragedy portrays either stark realism or a streak of cowardice in the next phase of Scully's horrible day. All the survivors of the Ballycohey massacre could walk, at least with assistance.

Scully said, "Come now, and let us hold a council of war."

"Well, what is best to be done?" asked Constable Cleary, the officer in charge.

"Let us at once make our way to the station," Scully suggested.

"No, I remain with my comrades till help comes. You have your own guard. Go if you will," said Cleary.

Scully objected, "What protection can you give a dead man; come and protect a live man." In any case, Cleary went part of the way with Scully and his guard to Ballykisteen House, just across the railway track in the direction of Limerick Junction. There, in the home of Jasper Bolton, Esq., the wounded were made comfortable, and medical help was sent for.

About that time, constable Hughes and another constable from the Monard station, located just past Limerick Junction, arrived with help. All that was needed then was an investigation for evidence. A short time later the subinspector, Mr. Saville, appeared, and one arrest was made. Morrow's body was placed in the house. Gorman was visited by Dr. Nadin, who found that he had been wounded in five places. He was given some medication and some water and was placed on a bed, where he expired at about six o'clock.

In the meantime Drs. Hewitt and Nadin promptly attended to Scully's wounds in the privacy of Ballykisteen House. All but one of the slugs were successfully extracted. By 7 P.M. he was back at his residence of Ballinaclough, escorted by a large constabulary force.

Early on Saturday morning, about 1:30 A.M., numerous constables from stations at Tipperary, Monard, Glenbane, Shanballymore, and Greenbane arrived at Ballycohey to scour the countryside for evidence. Eight additional arrests were made, but little evidence was gathered. Almost nothing was obtained in conversations with Ballycohey residents. The London *Times* expected that because of lack of local cooperation, little evidence would be forthcoming to solve the murders; but one of the police had been slain, so much effort was put forward to sift all available evidence. Unknowingly, the constables' investigation was better than they might have expected—three of the nine who were arrested were later acknowledged to have been participants in the Ballycohey shoot-out, and a brother of one of those who were arrested turned out to be the hero of the battle.[45] But hard evidence was lacking; there was never anything sufficient to bring anyone to trial, and all the prisoners were released.

Police reports in Dublin Castle are sketchy, since only the index to the Ballycohey outrage can be found. This brief source provides information on the murders, the arrests, a £500 public and a £300 private reward, and a sketch of the houses at Ballycohey. There were reports in the file at one time about further evictions, about protection for William Scully, and about the extra police tax assessed against the Ballycohey townland for eight additional policemen at four nearby stations. The last date in that file,

September 1, 1880, shows that the papers were returned to Dublin Castle by the chief secretary, but almost a hundred years later they cannot be found.[46]

Out-of-the-ordinary murders attract more attention than perhaps almost any form of news. Stories of the "Ballycohey Outrage" quickly spread through the British Isles. The London *Times* provided much coverage for the "tragic affray in Tipperary," generally treating it as an exception to the normally good landlord-tenant relations of recent years. The incident was depicted as the most "desperate and sanguinary" agrarian outrage in southern Ireland for some time, and the premeditation of the murderers at the "fatal ambuscade" was shown. It was noted that the usual reasons for such agrarian violence—namely, absenteeism and confiscation—were missing. Critical of all those who boldly defy the law, the *Times*, as its editors learned more of Ballycohey affairs, also criticized the landlord. He was regarded as "a very rash man; but every reader will be naturally disposed in favour of a man who, in the prosecution of his right, enters his own property, and presents himself as an easy mark to armed men denying him his right, and holding his property against him." Scully's contribution to the crisis, by introducing a new lease for Ballycohey, was described by the *Times* as "certainly an outrage on all reason, equity, and good feeling." The *Times* quickly obtained a copy of the lease and included it in their coverage.

Unlike many of the Irish newspapers, the *Times* did not hold the existing landlord-tenant law responsible. It continued to find in conditions surrounding Ballycohey an "exceptional, personal, and private character," based upon the actions of a zealous landlord and an equally determined tenantry. On one point the *Times* did have a complaint about the reasonableness of existing law. It could not account for "the spectacle of a man like Mr. Scully being able to obtain an escort of policemen to protect him in any high-handed and unreasonable proceedings he may wish to carry out." The *Times*'s evaluation is rather typical of an editor with prejudices. It carried the idea that its great distance from London made the Tipperary outrage of little value.

The day after the Ballycohey outrage, an inquest was quickly assembled in the courthouse in Tipperary town. Seven magistrates occupied their seats at the bench with the coroner, and a fourteen-man inquest jury was empaneled. After viewing the body of subconstable Morrow at the barracks and of bailiff Gorman at the workhouse, the jury listened to the evidence. The sessional solicitor of the county, George Bolton, Esq., of Nenagh, conducted the investigation on behalf of the crown. The verdict of this jury came quickly:

> That the jury find that Sub Constable Samuel Morrow and Darby Gorman came by their deaths on the lands of Ballycohey, from the effects of bullet-wounds inflicted on Friday, the 14th August, 1868, by some person or persons to us at present unknown. The jury are further of opinion that the conduct of Mr. William Scully, as regards proceedings toward his tenants at Ballycohey, is much to be deprecated, and the sooner legislative enactments be passed to put a stop to any such proceedings, the better for the peace and welfare of the country.[47]

The magisterial investigation, held privately after the inquest, dealt with those arrested on suspicion. Most of the prisoners were discharged for lack of evidence. It was also learned that "Dwyer, in whose house the fatal affray took place, was in the town of Tipperary during the whole of that day."[48]

A proclamation by the lord lieutenant of Ireland, offering a reward of £500, was posted throughout the counties of Limerick and Tipperary. "Any person . . . who shall, within six months, give such information as shall lead to the arrest of the person or persons who committed the outrage, shall be entitled to that sum. A further reward of Three Hundred Pounds will be given, . . . to any person who shall 'give such private information as shall lead to the arrest of any of the perpetrators.'"[49]

Most of William Scully's wounds healed quickly, but one gave him serious trouble. On the same day that he was wounded, the doctors were able to extract one of two bullets from his face; the one in his left jaw was not removed. Reports indicated other wounds on the fleshy part of his neck—one near the back part of an ear, and one over the temple. The first published accounts disclosed that Scully's wounds were severe but not likely to be fatal. Several days later the *Times* noted: "Mr. Scully is rather worse. One ball or slug still remains in his neck, and it is feared is working its way downward in the direction of the left lung." On the following day the *Times,* in more detail, stated that Scully's surgeons had extracted a second ball or slug from his jaw, but that they could not get the one in his neck. They reported that Scully had "manifested great impatience at the inability of the doctors to extract it, and called upon them to 'tear it out'—an expression which only indicated the acuteness of the pain he was suffering."[50] Two weeks after he was wounded, the news stated that "Scully is so far recovered, although the third slug has not been extracted, that he is able to go about the grounds, and, it is said, still carries his breech-loading rifle [shotgun]."[51] Surgeons were never able to remove that final bullet, and Scully carried that remembrance of Ballycohey to his grave.[52] On the whole, his wounds healed very quickly, and he was ready to carry out his plans for Ballycohey.

Because Scully was not wounded on the body, rumors spread rapidly that he wore chain mail under his clothing. A. M. Sullivan, in *New Ireland*, presented this story as a discussion at Cahir when one Irishman responded to another, "Arrah! how could the villain be killed when he wore *a helmet on his stomach*!"[53] Scully denied that he ever wore any kind of armor, and such denials were passed down to his family.[54]

In the meantime, William Scully's oldest surviving brother, Vincent, as head of the family, did not approve of the manner in which the stories about his youngest brother were being presented. With no apparent sympathy for William, Vincent, in a vain letter to the editor of the *Times*, wrote:

> Sir,—Observing that in most of those leading articles respecting the recent "outrage in Tipperary" which have appeared in *The Times* and other journals the name "Mr. Scully" is given without any distinctive prefix. I feel it right to remind you that a mere surname (especially in Ireland) would seem to designate the head of the family, and, consequently, in this instance,
>
> Your humble servant,
>
> Merrion-square, Dublin, Aug. 19 Vincent Scully[55]

William Scully, also concerned with his image, wrote a letter to an editor of the *Daily Express*, which was reprinted in the *Times*, saying:

> Statements have appeared in many papers, and some have been copied into your journal, characterizing my dealings, conduct, &c., as being illegal, tyrannous, and using other ugly words; and many unfounded statements have found their way into some papers. Time and space prevent me from saying more than contradicting the above, and asserting that my dealings and conduct have been not only entirely legal, but equitable and just. There is only one standard or rule to divide one man's right from another's. That rule is the code of civil law and equity of the land, and not Captain Rook's law. If I had transgressed that rule it was easy to get justice from the constituted courts. The murderers and their sympathizers knew this; but they also knew that the kind of rule they desired could not for one moment be sustained in reason or in law or equity. An appeal is, therefore, made to violence, intimidation, and unfounded statements.[56]

That was William Scully's valedictory comment to the press until very near the end of his life. That was about as long as the public and the news media could stay with a single story. Other news, particularly a serious wreck on an Irish mail train, soon displaced the Ballycohey outrage.

Scully's steadfast plans for Ballycohey were to evict all of the tenants, and he was more determined than ever because of events of August 14. On August 28 the police report stated that he had refused to "alter his course of treatment of his tenantry." Additional reports in September and October show that he requested and was granted police protection in order to deliver ejectment writs at Ballycohey.[57] After several more writs had been served, Charles Moore, from Mooresfort House, somewhat west of Ballycohey, appealed to Scully to sell rather than excite new violence: "Say what price you put on this Ballycohey property. *I will pay it to you,* and let there be an end to this dreadful episode!"[58] Moore knew nothing of Ballycohey's value, and he inquired of a tenant what rent they would be willing to pay. He learned the total rents and that tenants would go higher than they were paying Scully. With that information he negotiated with Ballycohey's landlord on a final price for the land, which is reputed to have been £14,500. There were additional points to work out before the sale was made with the transfer of the property on November 1, 1868.[59]

Scully gave up his police protection from October 30, 1868, through February 8, 1869, while he was absent from Tipperary, as things began to normalize; but the influence of Ballycohey did not die with the return of tranquility to that Tipperary farm. Irish historians generally attribute the passage of the Irish Land Act of 1870 to the extra attention given Irish landlord-tenant matters because of the crisis at Ballycohey. For many years, members of the Land League had been pushing for the Ulster customs of fair rent, fixity of tenure, and free sale of improvements to be extended to all of Ireland.[60] Prime Minister Gladstone was motivated to support the Land Act because of his understanding of landlord power in Ireland, as represented by Scully and others. While the new legislation provided more-complete legal protection for tenants, evictions did not end, but they became more costly to landlords, and the numbers of evictions reached a new low, compared to the previous twenty-five years. The law of 1870 also became basic to later land legislation in that it more fully defined landlord-tenant relations in Ireland.

The outrage at Ballycohey in 1868 inspired at least two literary attempts and produced numerous legendary accounts in Tipperary in later years. As might be expected, interpretations of Scully and his times there were grossly one-sided. "Rory of the Hill" became a popular ballad; it went like this:

I met a man on Slievenamon
    And I asked was Scully dead:
"I cannot give you that account,
    But I hear he's bad in bed.

He turned my mother out of doors,
  But I may meet him still;
I'm the bold Tipperary mountaineer,
  I'm Rory of the Hill."[61]

A poem with six stanzas by Tipperary poet J. J. Finnan in honor of Michael Dwyer (or O'Dwyer) after his death in 1902 makes mention of Scully, who had provided the situation that Dwyer seized to give him a hero image in his own land. The third, fourth, and fifth verses, which follow, have references that provide later-day Irish remembrances of William Scully in Tipperary:

Oh would that my voice could enliven the spirit
  That Irishmen true from their fathers inherit—
To love our old land and nobly stand by her
  With all the devotion and pluck of Dwyer
Who stood with his friends with the soul of a hero
  And guarded his home from the *black livered Nero*
Ignoring insults and old cherished opinions—
  And scattered or slew the *vile wretch and his minions*

You know what you owe to this man, sons of Erin—
  You know at the time how your heirs were despairing—
His deed like a thunderclap came to confound you
  And broke the first link in the *fetters* that bound you
Oh the blow that he struck can be still heard resounding
  And the ball that he fired shall keep ever rebounding
Till *wolves* from their victims are fated to sever—
  Till *despots* must go bag and baggage forever

Oh praise to the boys who stood shoulder to shoulder
  With gallant Dwyer—no man could be bolder
Oh bless their stout hearts, for they proved no backsliders
  But fought their own fight with no help from outsiders
If Irishmen all had their faith and reliance
  If Irishmen showed their bold front of defiance
Our land they could wrest from the big Saxon bully
  And smash him the same as these heroes *did Scully.*[62]

Of legendary quality are the accounts that say that Scully was driven out of Ireland by tenant resistance. Some, which are totally unrelated to chronological sequence, suggest that tenant resistance was the reason for his going to the United States, where he would have more freedom to require tenants to do his will.

By the time of the hundredth anniversary of the "Battle of Ballycohey," it was to be expected that Scully would be forgotten, or almost forgotten, in his native Tipperary. "Murderers" of one century became "heroes" of another. The accent by then was on "The Armed Resistance That Helped to Break Landlord Tyranny" or on the analogy that held that "Ballycohey was a perfect example of a just war." In the minds of some Irish patriots, Scully was so in league with the landlord class and the English government that it

The memorial to the "Defenders of Ballycohey," was erected on the occasion of the centennial observance of the Battle of Ballycohey. It is located in the Shronell Cemetery, about three and one-quarter miles west of Tipperary and about one and one-half miles from Ballycohey.

was convenient to forget that he was also a native-born son of Tipperary. Thus, when a memorial was erected in a cemetery near Ballycohey in 1968 to the memory of "The Fight against Landlordism," the fourteen men reputed to be the defenders at Ballycohey were shown as resisting eviction with firearms, which "So Frustrated the Despotism of Alien Landlords."[63]

Thus, by 1868, and even in 1968, Scully was treated as an alien on his own heath. But he retained all his other Irish landholdings after the Ballycohey affray, and he always kept a residence at Ballinaclough. However, his primary interests thereafter were elsewhere. By 1865, and certainly by 1870, Scully's wealth was increasing so rapidly that he could establish his residence virtually anywhere that he might wish. These changes of residence in the next few years and his extension of his holdings in the United States laid the basis for even greater wealth. Ireland, after Ballycohey, was merely a decreasing part of Scully's investments, and his periods of residence there diminished accordingly.

What of William Scully's vision for Ballycohey? A century later the entire area of this old farm was in lush pasture grass. Milk cows, birds, and small wild animals were the sole occupants. The sound of wind and animals was all that broke the silence. Scully was too far ahead of contemporary Irish agricultural thought; also he was too impatient and unfeeling toward occupants of the land. A century after the battle the Dwyer house was still standing, but was unoccupied; its last residents had moved only a short time before to the more exciting environment of Tipperary town. Likewise, Ballycohey was almost forgotten—its name no longer appeared on the official Tipperary map made by the national ordnance survey.

wswswswswswswswswswswswswswswswswswswswswswswswswswswswswswswswswswswswswswswswswswsws

5

Renewed Interest in America

wswswswswswswswswswswswswswswswswswswswswswswswswswswswswswswswswswswswswswswswswswsws

At Gurtnagap and Ballycohey, William Scully's efforts to rehabilitate his estates stirred up an active and murderous hostility among his tenants. He was readily identified as a "rack renter," who reflected adversely on other landlords, more because of the manner in which he carried out his landlord role than because of the rents he charged or the goal he had for his lands. Tenants banded together to oppose him; and their actions, even outside the law, received sympathetic support from Irishmen in general, and in many cases from Irish gentry. Such tenant opposition, especially when it was illegal, only heightened Scully's determination to maintain his legal rights. When his unyielding insistence on the letter of the law brought organized or armed resistance to his agents, who were only carrying out the landlord's directives, Scully became personally active and more determined than ever that his estates be handled in the manner that he prescribed. His unusual and extreme demands brought tenant retaliation, and the basis for the "crimes" at Ballycohey was often laid at Scully's feet.

No less contentious in setting his own pattern for his lands after he sold Ballycohey, William Scully did assume a less public—more private—guise in managing his landed estates. At the same time, he began to put idle capital into various types of securities, and in 1870 he initiated a vast expansion of his American holdings that was to continue for almost twenty-five years.

However, throughout the 1850s and 1860s, Scully's income came primarily from his Irish lands. After his bequest from Thomas in 1857, his net annual income from Irish rentals was about £3,000 sterling—which was equal to $15,000 to $18,000. The money coming in from sales of Illinois land, which had been initiated from 1855 to 1857, was not considered income by Scully. He treated it as capital, and he put that money back into other income-producing ventures.

In a summary of his annual income as of March 1 each year, Scully noted that his American lands produced no net income for the year prior to that date in 1866. Rentals of $4,300 were more than offset by taxes of $5,200. In the first fifteen years he was unable to make an annual profit on his American holdings. His patience in waiting for income from his American lands stands in marked contrast to his efforts to speedily gain greater income from his Irish estates. Scully was still dependent on returns from his Irish lands and from investments that he was making in England, where he earned £600 sterling that year. Nevertheless, his overall annual income was relatively high, between $15,000 and $20,000 for the year, which placed him very far up the scale in comparison with other incomes in either Great Britain or America.[1]

At the same time, expenses for operating his estates were relatively low, especially in Ireland. Scully paid £100 per year for the services of a full-time Irish steward, or bailiff, and other employees were paid much less. On the whole, no employees were an expense; they earned for the landlord more than he paid them. His American agents, who were employed on a part-time basis, earned almost as much as an Irish steward was paid. For Scully, 1866 would be the last year that his America lands would show a deficit. Thereafter, income from his Illinois holdings would increase rapidly, and he would look elsewhere enthusiastically for new income-producing properties.

During the next four years, Scully's old Irish estates continued to produce about £3,000 income each year. After 1866 he added income from Ballycohey, but he also had higher expenses, and he had difficulty in justifying the costs of the trials at Kilkenny. His summary of his income from sources other than his "old Irish estates" is shown in table 5.1.

On January 1, 1870, William Scully figured that his net worth was £212,700, which, if converted to dollars, would make him a dollar millionaire. Slightly more than half of that total was represented by 35,000 acres of land in Illinois, for which he gave an evaluation of $20 per acre in United States currency. To convert that sum to gold would take a 20 percent premium, so he figured that the Illinois land was worth $700,000, less $140,000, or in gold it was equal to $560,000, or £112,000. Additional wealth in the United States was represented by $197,000 in American bonds, on which he calculated a value of £34,100. At the same time the Irish estates left to him by Thomas had been improved to a value of £33,000, and the Irish lands that he had inherited from his mother in 1843 had been imroved slightly, to £27,000. Other wealth was represented by £4,400 "invested in cattle, sheep, horses, farm implem'ts and crops grown (chiefly hay) on all my Irish farms" and by £2,200 "invested in Russian Bonds, new ones."[2]

TABLE 5.1

WILLIAM SCULLY'S INCOME, MARCH 1, 1867, TO MARCH 1, 1870, EXCLUSIVE OF HIS "OLD IRISH ESTATES"

Year	Taxes and Improvements in Illinois	Net Rent Collected	Net Income from Other Sources (including Ballycohey)	Total Net Sum Received
1867	$5,170	$ 1,425	$ 5,000	$ 6,425
1868	7,010	1,875	9,000	10,875
1869	[higher than 1868]	5,100	11,000	16,100
1870	[higher than 1869]	25,000	7,000	32,000

SOURCE: "Net income from American lands, after paying for improvements on Ill.s. lands, & after paying Neb & Kansas taxes . . . ," in William Scully's handwriting, Office of the Scully Estates, Lincoln, Illinois.

William Scully's account book for bond purchases discloses that his early American bond acquisitions were United States government bonds, obtained August 26, 1867, at Fish & Hatch in New York City. Sizable purchases were also made from William Leese and through John Williams and Company. On March 1, 1870, the interest paid on these bonds for the previous year was $4,925. Earlier, Scully had purchased Orel-Vitebsk Railroad bonds in London, and in 1870 he bought a small number of Charkov-Krementschug Railroad bonds. On May 20, 1870, Scully paid $7,000 for Sangamon County, Illinois, bonds; and later that year, after he had sold most of his United States bonds, he purchased $153,639.57 worth of Chicago city bonds in order to generate further annual interest income.[3]

With resources of this kind, Scully began to look again to the American land frontier, as he had twenty years previously. By 1870 his Illinois lands were almost fully rented and were beginning to produce income each year that was as high or higher than the total cost of buying that property. The frontier for settlement had pushed westward along the steel bands of transcontinental railroads into the bluestem-grass country of central Kansas and Nebraska. No longer was there vacant land in Illinois, and trans-Mississippi states were rapidly filling up.

For many years, newspapers in Lincoln, Illinois, had reported the land sales in Kansas and Nebraska.[4] By 1869 more of these accounts of that far-western country had a personal connection, related to Logan County citizens who had traveled to or were planning to relocate in the new area. The Lincoln *Herald* reported: "After many 'hair breadth' escapes from the Indians, and from the exciting chases of the Buffalo, our two fellow citizens, Drs. Perry and Hunting have returned again to the peaceful quiet of home. They brought back no trophies in the shape of 'scalps and sich,' but each returned with a warrant for a section or so of Government land, and pronounce Nebraska as the 'land that flows with milk and honey.'"[5] Typically, Illinois farmers moved on or before March 1 if they were changing a location. So, early in 1870, when fifteen local farmers advertised farm sales, the Lincoln newspaper reported:

> PUBLIC SALES.—A western fever seems to have broken out in this community, and nearly every other man we meet has an attack of it. The words heading this, greet you on every side, turn where you will. For the information of our readers we will give a list of sales that will be made in the coming two or three weeks, and for which we have printed bills. Not all of the parties selling out are intending to emigrate, however, but the major portion of them have their eyes fixed on the lands beyond the Mississippi.[6]

No available source tells what prompted William Scully to take the next step leading to his acquisition of virgin land in Kansas and Nebraska. There is strong evidence of "westering fever" in Logan County and that Scully's Logan County agent, William McGalliard, was coming down with that urge to seek good cheap land in the west. Moreover, John Williams, a Springfield merchant and a long-time Scully confidant, acquired a large quantity of government land on the frontier of Kansas by May, 1870, and some of that property was eventually purchased by Scully.

Because of his recent sale of Ballycohey and because of the vastly increased revenues coming from the Scully estates in Illinois, it was evident that Scully could buy much land without going into debt or depending on bankers. Contemporary evidence about Scully's personal connection with the purchases at federal land offices located in Beatrice, Nebraska, and Junction City, Kansas, is scant. Certainly, he strongly endorsed the purchases, but Scully family stories recall that William Scully gave instructions on western land-buying that no land was to be acquired west of the Blue River tributary of the Kansas River. Because all the government land entered in both Nebraska and Kansas was somewhat west of the Blue River, the question arises, Who made the decision to alter the original instructions? Also, there are stories, reminiscent of accounts in Illinois, which say that land seekers did not see William Scully in line at the land office, yet the

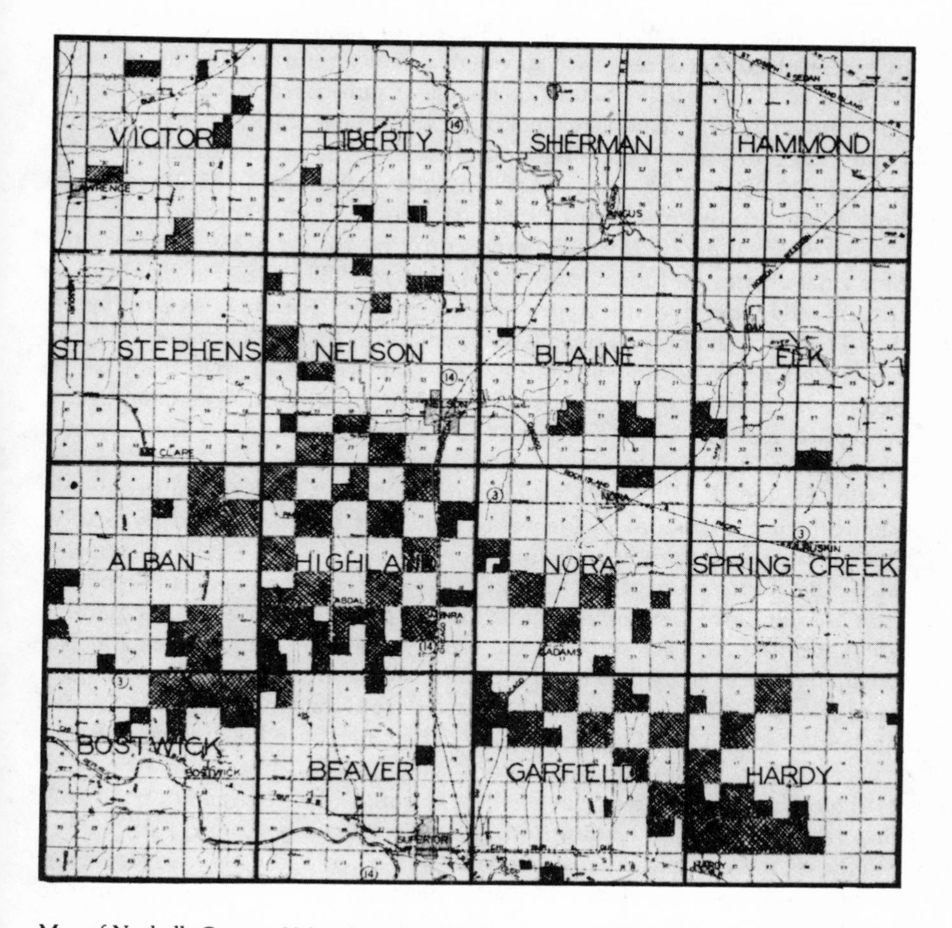

Map of Nuckolls County, Nebraska, showing land that Scully owned there in 1900, which he had purchased in 1870 at the Beatrice Land Office. Courtesy of the Nebraska Historical Society.

very land that they wanted was taken by Scully just before they had the opportunity to enter it for themselves. Still other accounts hold that Scully bought his position in line from a poor man who was willing to start over again at the end of the line. And again, there was the legendary recounting that William Scully carefully considered the land that he bought by digging hundreds of little holes in the prairie top soil with the same spade that he had used twenty years earlier in Illinois.

One author, in imaginative and picturesque fashion, wrote: "On June 13, 1870, William Scully walked into the Beatrice land office and calmly paid the Receiver of Public Monies $38,084 for a block of Government land in excess of 30,000 acres. Two days later Scully returned to buy an additional 6,400."[7] Actually, Scully or his agent purchased 40,686.97 acres of government land, located in Nuckolls County, Nebraska, at the Beatrice Land Office in June, 1870. These purchases, at the rate of $1.25 per acre, totaled $50,858.71. They were for ten tracts, equal to 2,785.21 acres, on June 6; seventy-one pieces, for 29,317.04 acres, on June 13; three pieces, for 320 acres, on June 14; thirteen parcels on June 15, for 6,239.88 acres; and nine final entries on June 22, for 2,024.84 acres.[8] The General Land Office took much longer this time to provide land patents—as much as three years.

Twenty residents of Logan County, Illinois, had "westering fever" and bought government land at the Beatrice Land Office in 1870 prior to the Scully entries. One of them was William McGalliard, Scully's chief agent, who purchased 640 acres on May 14. Interspersed among the Scully entries of June 15 and 22 are eleven other McGalliard purchases for 7,596.96 acres, costing $9,496.20. With his May 14 acquisition, McGalliard acquired nearly one-fifth of the amount of Nebraska land that was purchased by Scully. Such comparisons produce the question, How fully was Scully sponsoring McGalliard's investment in cheap government land? Later, agents reported that the landlord perennially gave advice on the procedures for gaining great wealth. At least some of the money that McGalliard spent on Nebraska land was loaned to him by Scully, for on March 1, 1871, this agent paid off a $3,000 note for money borrowed from the landlord and redeemed his land deed and Receiver's certificate, which had been held for security.[9]

Together, Scully and McGalliard accounted for more than 70 percent of the cash purchases made at Beatrice in June, 1870. It was not until 1871 that large-scale homesteading took place in Nuckolls County; so Scully's purchases there came before much land was taken in the county. However, because settlement was just under way, he could not expect much chance to rent his land in Nuckolls County for many years.

Cash purchasers of government land at Beatrice in June, 1870, were from states such as New York, Iowa, Pennsylvania, Connecticut, and Illinois, as well as nearby eastern Nebraska counties. Other large cash purchasers at Beatrice that year included Maggie C. Blakely, who must have been the wife of the receiver of the land office; Jacob Shoff, a land speculator from Otoe County, Nebraska; Jacob Livengood of Somerset County, Pennsylvania; Andrew Cropsey, a heavy investor in government lands, from Lancaster County, Nebraska; William E. Ide of Franklin County, Ohio; William Youle of Tazewell County, Illinois; Mahlon W. Keim of Cambria County, Pennsylvania; and various members of the Arnold family of Sangamon County, Illinois.[10]

One hundred miles south and a little west of Beatrice, at the Junction

City, Kansas, Federal Land Office, forty-eight entries for 14,610.55 acres of land, located mostly in Marion County, with several small parcels just across the line in Dickinson County, were purchased on behalf of William Scully on July 1, 1870. Acquired at the government minimum price, this land cost $18,263.19. McGalliard made no personal entries of land in Kan-

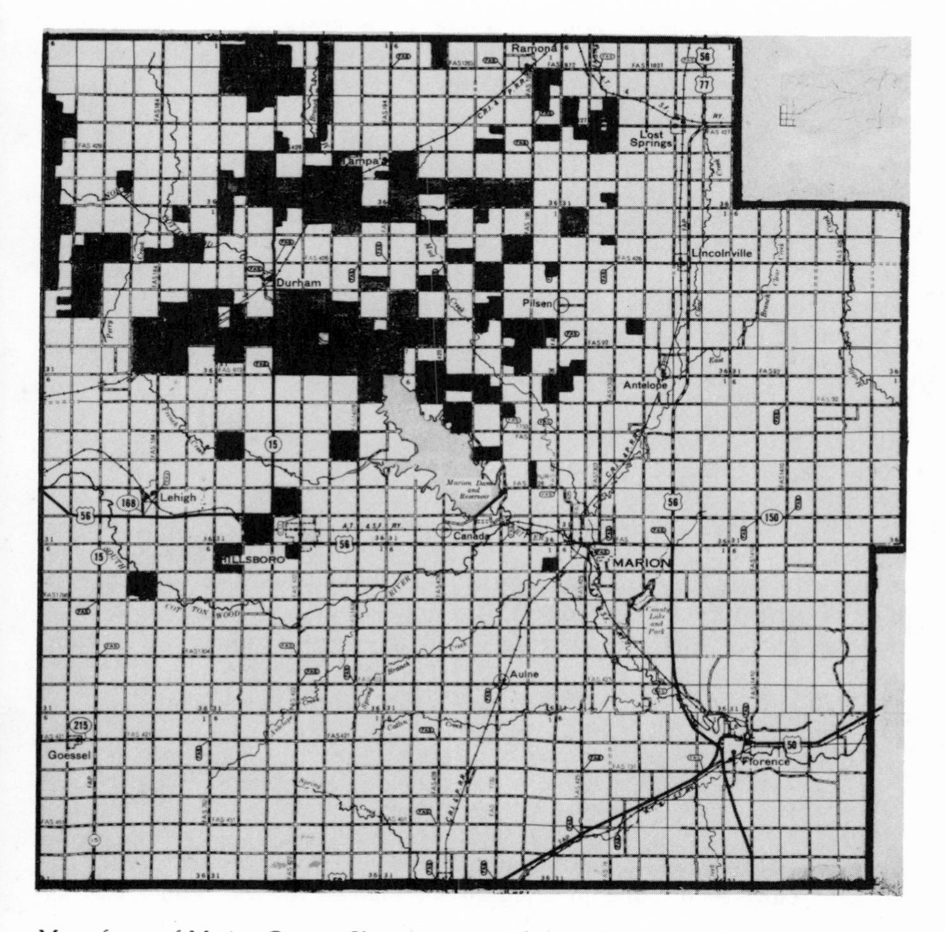

Map of part of Marion County, Kansas, in 1971, showing land that Scully owned there in 1900. Scully also owned 1,120 acres in Dickinson County, just north of Marion County. Scully heirs have subsequently sold about 4,000 acres of their land in Marion County to the federal government for use in the Marion Reservoir and to the city of Hillsboro for use as an airport.

sas. In this first purchase, care was taken to avoid locations within the area of the railroad's primary land grant, where the government selling price was $2.50 per acre.[11] Heavy purchasers at Junction City in 1870 included Walter B. Beebe of Harrison County, Ohio; T. J. Peter of Shawnee County, Kansas; John Williams, long known to Scully, of Sangamon County, Illinois; Francis Skiddy of New York City; and John Niccols of McLean County, Illinois.

After 1870 the days of large-scale purchase of public lands was just about over. The unrestricted lands available for private acquisition through cash purchase was nearly gone. In the next few years, William Scully obtained more than eight thousand acres of government land in Marion County, and except for about four hundred acres of double-minimum land at $2.50 an acre, he paid the regular price of $1.25 per acre. Almost all of this land had heavy, deep soil covered with perennial bluestem grasses and was suitable for cultivation; however, some would always remain pasture. If Scully were inclined to add to his estates in later years, he would have to pay the higher prices demanded by private owners. In the meantime, as he had done in Illinois, he appointed agents at Nelson, Nebraska, and at Marion, Kansas, to watch over his new holdings in Nuckolls County and in Marion and Dickinson counties.

Typical of these agents was Alex E. Case of Marion, who handled Scully's affairs in Kansas for many years. Case arrived in Marion Centre as a twenty-eight-year-old Pennsylvania Civil War veteran in 1866. He immediately became county surveyor, a post he held for twelve years; and he served as the elected county attorney and, in 1869, as a member of the lower house of the state legislature. In 1868 Case established his own insurance agency, and he began serving as land agent for the sale of extensive land grants owned by the Atchison, Topeka and Santa Fe Railroad. Case was the type of person that Scully sought as agent—one who was active in local political and civic affairs—and Case willingly served him on a part-time basis.

Extensive correspondence between Alex Case and Scully's office in Lincoln, Illinois, helps to explain why more subsequent purchases were made in Marion County so as to make it the banner county in number of acres owned anywhere by William Scully. These letters indicate that Scully had visited the county and that he was interested in adding to his holdings there. Specific land areas were identified that would help round out the already extensive Scully estate, and instructions were given for the proper way of preparing a completely valid and acceptable title to any land purchased by Scully.[12] Offers were made to out-of-state owners for some land.

Moreover, Scully's chief agents in Lincoln, Illinois, subscribed to Marion newspapers, which apprised them of lands that were available for sale there, as did letters sent to them by Case. Notes had obviously been taken regarding the availability of certain tracts; a map of Marion County was kept in the Lincoln office; and correspondence was quickly answered. Mail moved rapidly, and usually within a week, Case would get a reply to an offer made on additional land. If an acceptance was not forthcoming, Case usually asked Scully's agent, "What am I bid for that land?" Land prices in Marion County were particularly low, due to the effects of the Panic of 1873 and to the distress of local landowners from the grasshopper plague of 1874; so, many of these prices quoted to Scully were rock bottom.

An example of the results of these negotiations by the Marion County real estate agent was Scully's purchase in 1876 of three sections, 1,920 acres, for a price of less than $3.00 per acre. The sellers were Louis Tuckerman and Morton Redmond, who had acquired the land six years earlier for just over $2.10 per acre. They had gotten the land from a short-term speculator, who had obtained it from a speculator, who had made the most money because he had acquired these sections in 1869 from the government at $1.25 per acre. Much time and correspondence was spent on this sale, and Tuckerman and Redmond stood the cost of selling the land, including a 5 percent commission to Case and payment for the abstract of title and recording of the deed. Their profit from six years of ownership was modest by any standards, but no doubt they were happy to exchange the land for cash.

In this and in most other Scully purchases in Marion County, Case, as an independent real estate agent, served the seller of the land; and at the same time he was trusted by Scully to provide a title that was free from imperfections of any kind. For example, even though an abstract of title did not note a failure to pay taxes in 1874 on land that Scully purchased in 1876, the landlord discovered the omission, and the penalty and back taxes were paid by the seller at Case's urging. Scully also bought from John Williams, his good friend and long-time acquaintance in Springfield, Illinois, 9,440 acres in Marion County at a cost of about two dollars an acre. From the Santa Fe land grant in Marion County he obtained 8,622.46 acres in five separate transactions, for a cost of slightly more than $4.50 per acre.[13]

Case, the Marion real estate agent, sought out owners of land in the county who were willing to sell to Scully. Some lived far distant from Kansas, in New England, Canada, or California. An examination of the deeds of Scully's Marion County properties shows that his purchases there averaged three-quarters of a section in size and that most tracts were somewhat improved. Most of this land had had no more than four or five owners since the patent had been obtained from the federal government. True, Scully made large purchases, such as those from Williams or from the Santa Fe, but he bought from large and small sellers alike, and his terms for providing a legal sale were fully approved by Case. In response to one seller, Case replied: "We are in receipt of your recent letter. Mr. Knox has *not* sold Mr. Scully your land and *we have,* and you can have your money

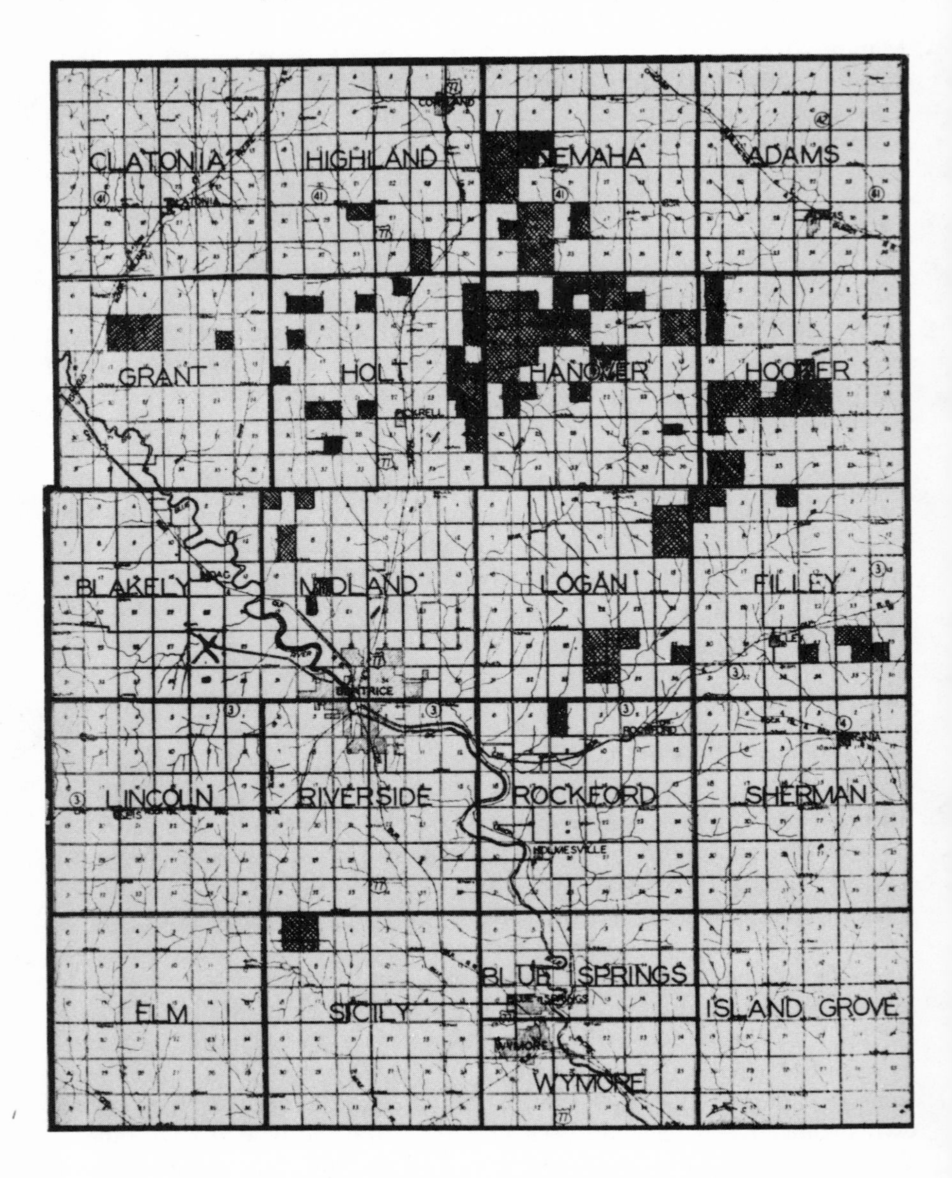

Map of Gage County, Nebraska, showing land that Scully owned there in 1900, all of which he had purchased from private owners. Courtesy of the Nebraska State Historical Society.

as soon as you comply with the conditions of the sale. Mr. Scully is not that kind of a man; he will not get back on an agent in that way we assure you." Five days later, Case wrote again: "We have the money from Mr. Scully to pay for your land on receipt of deed and deed abstract of title showing title perfect in you. Mr. Scully wants this transaction closed at once as he has other lands offered him which he wants unless he takes this."[14] Case's direct manner must have appealed to Scully, for they had business dealings for many years. When the time came for a full-time Scully agent in Kansas, Case was offered the position, which he declined because of his many other business commitments.

No additional land was acquired by William Scully in either Nuckolls County, Nebraska, or Dickinson County, Kansas. But sizable purchases were made in the early 1880s in Gage County, Nebraska, some fifty miles west of the Missouri River; in Marshall County, Kansas, just south across the border from Gage County; and in Butler County, just south of Marion County. Most of Scully's acquisitions in Gage County were north or east of Beatrice—east of the Blue River—with the largest concentration in Hanover Township. Except for two sections, originally acquired by the state of Nebraska, all of the Scully land in Hanover Township was entered with land scrip made available through the Morrill Land Grant College Act of 1862. By the time of Scully's first purchases there in 1881, the land sold for prices ranging from $6.25 to $7.50 per acre, with rapid increases in later years.[15]

Scully's wealth was increasing very rapidly in the 1870s and 1880s, and this growth pattern did not seem to be hampered by downturns in the economic cycle, such as the one brought on by the Panic of 1873. Scully prepared a chart (table 5.2) of his net income and added to it through early 1879. This chart shows something of the landlord's financial condition in the 1870s.[16]

On April 3, 1879, William Scully wrote: "I have saved for the last 9 years from March 1/70 to March 1/79 just about $70,000 per year. Total $630,000 and I still have it, in the form of either Lands, or good Bonds, Safe."[17] This was the money that Scully used in making new land purchases for which he paid much higher prices than he did for his initial land investments. For instance, he started buying land in Illinois again in 1875 by acquiring 824 acres in Logan County and 640 acres in Livingston County. In 1877, in the midst of depression times, 263 acres in Logan County and the huge Hamilton tract in Sangamon County, 4,161 acres, were picked up. In 1878, 1879, and 1880, Logan County purchases were 1,662, 751, and 160 acres respectively. In 1883, 1885, and 1886, smaller purchases were made in Logan County: 80, 188, and 114 acres. A final Illinois purchase was 1,500 acres in Livingston County in 1887. The

TABLE 5.2
WILLIAM SCULLY'S INCOME FROM MARCH 1, 1871, TO MARCH 1, 1879, EXCLUSIVE OF HIS "OLD IRISH ESTATES"

Year	Taxes and Improvements (including land purchases) in Illinois, Kansas, and Nebraska	Net Rent Collected	Net Income from Other Sources	Total Net Sum Received
1871	$------*	$36,000	$12,630	$48,630
1872	------	35,240	13,380	48,620
1873	------	36,750	15,280	52,030
1874†	------	37,700	30,046	67,746
1875	16,200	54,000	31,300	85,300
1876	14,385	45,618	32,300	77,918
1877	17,783	57,850	32,300	90,150
1878	14,737	49,547	31,670	81,217
1879	------	------	20,370	------

* The dashes indicate that no figures were available.

† A "year of great loss" was noted, no doubt a reference to the grasshopper plague of 1874, which was destructive to summer crops in Kansas and Nebraska.

figures that Scully set down in 1888 for his United States lands are shown in table 5.3.

In comparison with the systematic handling of the Scully lands in later days, the first twenty to twenty-five years in Illinois seem to have been unplanned and chaotic. After the initial land purchase there in 1850, John Williams was asked to help see that taxes were paid promptly. After the Scully acreage built up in Logan County and after the abortive effort of the landowner to live on his land and to develop it on his own, Samuel C. Parks was appointed as Scully's legal representative for overseeing the lands during the owner's extended absences.

With the growth of Scully's leasing activities in Illinois, William McGalliard became Parks's partner; then when Parks left for the West, where he eventually served on the Territorial Supreme Court of Wyoming, McGalliard became Scully's leading agent in America. McGalliard was apparently influential enough to get Scully interested in the trans-Missouri-River country where he more than doubled his land acreage in 1870, thus laying the basis for an even larger estate.

McGalliard was a native of New Jersey who had arrived in Lincoln, Illinois, in 1858. A lawyer, he was active in the Union and the Republican parties, and in both 1866 and 1867 he was elected to the Illinois state legislature. He also served a term in the early 1870s as master in chancery. After a number of apparently successful years as Scully's agent, McGalliard

TABLE 5.3
WILLIAM SCULLY'S UNITED STATES LANDS, OCTOBER 22, 1888

Location	Acres	Initial Cost	Cost per Acre
Illinois, by County			
Originally entered & bought: Logan	29,686		
Tazewell	601		
Grundy	8,412		
Will	400		
Mason	10		
	39,109		
Sold in 1850s	4,280		
Original land on hand, October 22, 1888	34,829	$ 35,000	$ 1.00
Bought from private sellers in Logan, Sangamon, and Livingston, 1875–87	12,328	523,417	42.46
Totals for Illinois	47,157	$ 558,417	
Kansas, by County			
Originally entered & bought: Marion and Dickinson	23,193	$ 29,500	$ 1.27
Bought from private sellers: Marion	31,284	144,106	4.61
Marshall	8,215	85,956	10.46
Butler	9,058	86,910	9.59
Totals for Kansas	71,750	$ 346,472	
Nebraska, by County			
Originally entered & bought: Nuckolls	40,830	$ 51,038	$ 1.25
Bought from private sellers: Gage	23,157	292,785	12.64
Totals for Nebraska	63,987	$ 343,823	
Grand totals for U.S. land, 1888	182,894	$1,248,712	$ 6.83

SOURCE: Memorandum in Scully's handwriting, dated October 22, 1888, in the office of the Scully estates, Lincoln, Illinois. Scully's total here is several hundred acres larger than the figure derived by adding the Scully acreage from each county and each state.

"shot himself at his rooms over Mayfield's bank" and died on November 11, 1873.[18] No motive was ever given for this suicide. There were some arrears in rent collections, but the amounts were not large. Interestingly, Scully did not acquire McGalliard's western lands from his former agent's estate.

In December, 1873, Scully appointed a Captain F. Fisk as his agent for a short time. Then, early the next year, new agents were selected, and a new phase in the administration of the huge William Scully estates in the United States began.

wswswswswswswswswswswswswswswswswswswswswswswswswswswswswswswswswswswswswswswswswswsws

6

Order to Scully Business

wswswswswswswswswswswswswswswswswswswswswswswswswswswswswswswswswswswswswswswswswsws

The continuity of Scully agents in Illinois was broken with the death of William McGalliard late in 1873. It did not take the landlord long to decide what should be done. After the Lincoln office for the Scully estates had been guided by Captain Fisk briefly in a transition period of only a few months, the firm of Koehnle and Scully was created early in 1874 to handle the business of William Scully in the United States.

Frederick C. W. Koehnle, the older member of the partnership, was thirty-eight years of age when he became a Scully agent. He was born in Germany and had been in the United States for twenty years. A resident of Logan County since 1855, Koehnle was very active politically following his election in 1860 as a trustee for the town of Lincoln. In 1862 Koehnle's Addition was platted into new lots for Lincoln. He became school inspector in July, 1867, alderman in 1869, and for two terms he served as clerk of the Logan County circuit court. Koehnle listed his politics as Republican, and he attended the Liberal Republican National Convention in Cincinnati in 1872. His religious persuasion was Evangelical Lutheran, and he was an officer in his congregation.[1]

The Scully member of the firm was John Scully, born August 30, 1849, in Dublin, Ireland. Neither William nor John ever acknowledged a close kinship; they were father and son, although the mother of this illegitimate union cannot be identified. Newspapers often assumed a father-son relationship, just as they frequently identified John as a nephew of William's. Presumably the mother was Irish: thus John was three-fourths Irish, whereas William, who had an English mother, was only half Irish. Unconfirmed stories from later years state that John's mother married, that she came to the United States with her husband, and that they became Scully tenants in Logan County, where they failed to pay rent—at least at times—without being threatened with eviction by a landlord who was not hesitant to use that legal weapon on others.[2] If that was true, they never disclosed to others a past connection with William Scully.

It is believed that William Scully never formally adopted John, but that he took a close personal interest in his welfare, paid for his education, and was proud of his scholarly achievements. John's preparatory schooling was in England, where in 1863, at the age of fourteen, he entered the prestigious Rugby School, famous nationally after the publication in 1857 of *Tom Brown's School Days* by a former Rugby student. Chartered in 1567 and fully endowed by 1653, Rugby School, located in the English midlands just east of Coventry, did not attain its scholarly distinctions until early in the nineteenth century. The headmaster during John's four years there served as archbishop of Canterbury in later years.

Although John's examinations and his tutorial reports won several prizes at Rugby, he was not sent to a university. Instead, William helped to place him in an apprenticeship for three years with a London law firm. This arrangement was extended for two years, and at the age of twenty-three, William offered him the opportunity to serve as his agent in Lincoln, Illinois.

The partnership of Koehnle and Scully, which inverted the firm's name to Scully and Koehnle two years later, provided mutually advantageous skills to the landlord. Koehnle knew Logan County well and understood the temper and attitudes of its citizens. John Scully had a long acquaintance with William Scully, and he was trusted more fully than were other agents. Both partners were hard working, vigorous, and highly motivated. They were able to take William Scully's directions, some given orally and some provided in detailed letters, and transfer them into procedures for providing a profitable annual income from the extensive land areas spread out in three different states. There were no immediate plans for extensive record- keeping: that would come later. But the partnership was exceedingly active in adding new acreage to Scully's holdings, as noted in the previous chapter.

It continued to use the printed one-page leases—fairly typical of leases found on other Illinois farms—that had been used for renting Scully land for a number of years under McGalliard's direction.[3] Record books set up by McGalliard for recording each tract of land and the nature of the rental agreement were continued. There were also several other books for recording special accounts, but on the whole the growing Scully enterprise was operated with few written records and much personal contact between agents and tenants. Two pages from William Scully's rent accounts in

Logan County, covering nine years for one tenant farm, illustrate the simplicity of these early records (see table 6.1).[4]

TABLE 6.1
RENT ACCOUNT ON ONE SCULLY FARM, 1871–79

	Date		Rent	Tax	Total
1870	Jan 1, 1871	Alexander Miller	$120	$21.14	$141.15
	N½SE¼ 23-21-3	80 acres			
1871	Jan 1, 1872		"	23.05	143.05
1872	Jan 1, 1873		"	17.63	137.63
1873	Jan 1, 1874	Fred Zurkammer	200	19.66	200.00
				balance 19.66 pd 2-18-74	
1874	Jan 1, 1875		"	18.45	200.00
				balance 18.45 pd 2-6-75	
1875	Jan 1, 1876		"	25.15	138.75
				86.40 balance pd 3-2-76	
1876	Jan 1, 1877		240	16.57	302.80
			40.17 balance		40.17 carried over
1877	Jan 1, 1878		220	24.96	balance $285.13
		balance 40.17			
1878	Jan 1, 1879		220	24.99	pd $100 5-11-78
		balance 285.13			108.75 1-31-79
				balance $321.37	

The years 1874 and 1878 were difficult ones for farmers in Logan County, and many of the Scully tenants received abatements on their assigned rent during those years that totaled half or more of the amount of their rent. When the landlord forgave rent in this manner, it was sometimes recorded in the rent book with a reason, such as "bad crop" or "poor crop & wet land."

The construction and maintenance of fences was a more difficult problem for farmers on unforested land, such as Scully owned in central Illinois or in Nebraska and Kansas. In later years the solution was to use barbed wire. But in Illinois in the early years a proper living hedge fence would serve the purpose. So, another record maintained for Illinois Scully lands was a "Hedge Book," in which notes were kept on contracts with tenants for setting out and maintaining hedge fences. Thus, it was recorded on September 13, 1870, that John and Thomas Garrett had leased 160 acres and that "hedge being made prior to this lease [the] Garretts were compelled (on back of lease) to keep in good order & preserve & keep the hedge properly trimmed down, as a provision of the lease." According to Scully's agreement, the hedge plants were provided and the tenants were paid to plant them and properly care for them until they were good live fences. When a tenant failed to comply with his part of the bargain, Scully would not pay the agreed price. For instance, when the agent checked Fred Frorer's lease after he had entered into a hedge contract on October 23, 1874, he found that "hedge examined—N. line very poor, never cultivated—W. line poor, a great many gaps plowed once in fall—full of weeds.—to be plowed up entirely, or to be cut off the ground & reset.—No pay." In order to get the tenants to comply, many of the hedge agreements included in five-year leases in Logan County stipulated that rates of from 35¢ to 50¢ and even 60¢ per rod of hedge were to be paid at the end of the leasing period.[5] Typically, the landlord also provided fruit trees for orchards and bricks for foundations or for lining wells.

The printed Scully lease by late 1876 was beginning to take on a longer, more-detailed characteristic, which was peculiar to leases in later years. This lease, which more often specified cash rent and certain requirements to be fulfilled at explicit times, was sent to agent Alex Case in Marion, Kansas. He responded by saying: "It looks all right and terms easy which is also favorable if you have plenty of lease blanks on hand. Wouldn't it be well to forward us a supply[?]"[6] A few pieces of Scully land in Marion County might have been leased by that time, but leasing picked up after that, slowly at first, then rapidly as adjoining land was fully settled and placed under cultivation.

Most of these Kansas leases of the late 1870s and early 1880s were five-year contracts, although one- and three-year leases also existed. When there were no improvements on the land, the tenant agreed when signing a farming lease to pay for his first two years' rent by breaking a certain amount of land and also to pay the amount owed in taxes on the land. On a quarter-section lease a typical rent in the third year was $50 plus taxes; fourth year, $115 plus taxes; and fifth year, $160 plus taxes. The house and farm buildings on a Scully lease were provided by and owned by the tenant, whereas buildings on most rented farm land belonged to the landowner. By 1883, forty-five Scully leases were in force in Marion County, fewer than 10 percent of the number some ten years later. In Nuckolls County, rents increased from slightly more than four thousand dollars in 1889 to ten times that amount in 1899, a suggestion that leasing in that far-western county came at a slightly later date.

One-year leases of the early 1880s in Marion County were for haying, for which the tenant paid 25¢ per ton of hay taken from the land. As a conservation measure, later Scully leases would specifically prohibit removal of hay from a farm; however, a tenant could use the hay for his own livestock. Some of the three-year leases were for farming, and some for grazing, for which nominal rates were charged. When a lease was made on an improved farm, or when leases were renewed, rental payments plus taxes began from the first year on the lease.[7] One-year leases were more typical at this time on Illinois land. Most of these annual leases were renewed almost automatically to the same tenant. By the 1890s all of the Marion County Scully land was under lease, a lag of about the same magnitude as that encountered earlier in Illinois.

But the delay in Marion County and in other western lands was not due to the swampy or boggy character of the soil. That perplexing question was unique to the Illinois portion of Scully's American lands, and the landlord, operating through his full-time agents, the firm of Scully and Koehnle, sought to do something about the problems of the old Scully Swamp and to make the land live up to its productive potential.

William Scully had direct experience in Ireland with the value of proper drainage. He understood the principles of drainage, and he had seen the dramatic results for grass- and crop-production through eliminating problems of wet land. But Scully did not use his expertise in advance of other draining efforts in various parts of Illinois. Much of his land was concentrated in one large block, which would prove to be an advantage when the time came to conduct massive drainage operations. Thus, Scully could develop his drainage patterns to fit the contours of his land, and he would not have to await approval or cooperation of adjacent landowners to initiate his own program. The scant information that is available shows that some programs for drainage of Scully lands came before the passage, in 1879, of an Illinois law providing statutory support for organizing semigovernmental drainage districts.[8]

Scully lease records, as early as 1870, show that the landlord was allowing a tenant half of the cost of cutting a ditch, presumably needed to drain the land.[9] Later the landlord would assume the entire expense of such efforts, in agreement with his long-time assumption that the landlord owned the land and the improvements in the soil, such as wells, trees, foundations, and drainage tiles, whereas the tenant owned the above-ground improvements, such as fences and buildings.

In his income and expense accounts of the 1870s, William Scully first listed a separate expenditure for ditching in 1875. From that small beginning, for almost forty years there were new drainage projects being developed for portions of the Illinois holdings.[10] There may have been small hand-dug ditches made without landlord or agent supervision prior to McGalliard's death, for which the landlord did not reimburse the tenant. Legendary accounts exist of the early use of an implement called the mole on Scully's Logan County land. The working part of the mole was a bullet-shaped object attached to a plow in place of the moldboard. When a team of horses pulled this implement, with the mole a foot or more underground, it would open up a tubelike hole from four to six inches in diameter.[11]

Although these holes would collapse in time, they could provide much drainage of surface and subsurface water.

Lewis Jefferson ("Jeff") Sims, former sheriff of Logan County, who had no experience in draining wet lands, was selected by Scully to oversee tiling and ditching of Scully Swamp. He was sent by the landlord to the University of Illinois at Urbana to take a short course in drainage engineering. When he came back as a $100-per-month foreman of a tiling crew of fourteen to seventeen permanent tilers on Scully lands, drainage efforts took on an active and aggressive stance.[12] Many tenants and day laborers also worked on these projects on a seasonal basis.

Specific techniques and distinctive equipment were used to lay in drain tiles, which were purchased in railroad boxcar lots. Once a year, Scully agents bought a carload of four- to fifteen-inch tiles from the National Drain Tile Company or the Joliet Mound Drain Tile Company or some similar firm. Tenants were hired to transport tiles from the railroad siding to the work sites, something that could be done in the off-season for heavy farm work. Sims and his crew took over from there.

As a long-time Scully employee, Jeff Sims became expert in using a transit to get proper drainage for each line. He planned each section of this vast drainage network, basing it on soil samples and a study of the surface of the land. Tiles were placed deep enough to remain undisturbed by farming operations, and each section of tile was built large enough to handle potential water demands. Sometimes, work on a particular piece of ground would be laid out in three stages, to be completed in three years. During the first year only the large main-line ditches would be dug and the tile laid; the second year, branch lines would be added; and the third year, tiles would be entended to higher points in the field. The hardest work of the tiling crew was digging ditches by hand with tiling spades, although horses pulling a slip were used for the larger ditches. A half-size wagon, built especially for use on soft ground, was drawn by a single pony to carry the tiles to the ditches.[13]

Larger open ditches and canals to serve as outlets for the drainage tiles were dug out by a floating dredge. This large-scale activity was the most impressive of all of the Scully draining operations in central Illinois, and it was long remembered by area residents. A dredging crew generally had with it a barge, containing a building two stories high, which was their sleeping and eating quarters. Large equipment of a mobile nature, such as the dredge and the quarters for the crew, seemed foreign to the flat lands of the drying-up Scully Swamp, and the oddity of this activity attracted much attention.[14]

Not until after 1910 was the mammoth operation of laying in the tiles and dredging out ditches and canals on Scully's Illinois lands completed. Thereafter, a much smaller part-time maintenance crew inspected the drainage works and made necessary repairs. This immense system cost much more than William Scully had expected—more than a quarter of a million dollars—but it greatly increased the capacity of his land to produce in all kinds of weather.[15]

Although William Scully was convinced of the value of draining his wet lands, he sought assurance that his expenditures of so much money on getting his drainage system built was advisable. During many of the years of the 1880s and 1890s, 10 to 15 percent of the rental income from the Illinois land went back to the soil in new drainage works. Typically, the landlord questioned his agents about expense of this magnitude. Finally, in 1892, they answered at length to justify this heavy outlay. Their response used a Logan County quarter-section as an example of how much was spent and what increased revenue could be expected. Sixteen thousand feet of tile were laid in this single quarter-section in the summer of 1887, at a cost of $334 for the tiles and $521 for labor and other expenses. Rents on that quarter varied from nothing in 1877 to an unusually large $1,051.12 in 1881, "when the tenant was sold out & everything taken." The average for ten years, 1877 to 1886, had been $428.83. The agents further explained that when the tiling was completed in 1887, "this land . . . was in such horrid condition, from bad farming in wet ground & full of weeds that we had to lease it for 5 years to Gerd. Harms, a very energetic farmer," for rents gradually increasing from $498.33 in 1887 to $545 in 1891, a five-year average of $524 per year. After 1892, rents were to go to $600 per year, an annual difference of $172 from ten years prior to the tiling.[16] A similar justification was made for installing drain tiles in Sangamon County, where expenditures of $9,500 over a number of years were offset by an annual increase in rental income of $1,400.[17] Farming was so much better on drained land that tenants were not expected to have any difficulty in paying the new higher rates.

The magnitude of the Scully draining projects drew visitors from the local area and from distant points to see how the tiling was being done so they could apply it to their own wet lands. Tiling records, kept in the Scully Estates Office, show that 135,117 feet of drain tile was used on a single section of land—number 10-20-4—in Logan County. Most of this tile was four inches in diameter (95,931 feet), but it ranged upward to fifteen-inch tile (2,859 feet). With forty-eight sections of land in Logan County, Scully had a potential of 1,200 miles of drain tile in that county alone if drainage efforts were equally developed.[18]

With the completion of Scully's vast drainage system, certain conservation practices were incorporated into leases. Banks along ditches and streams were vulnerable to erosion, so tenants were prohibited from farming within ten feet—later the distance was increased to fifteen feet—of the ditches. Grass was to be maintained on the sod berm on the banks and along waterways that would hold back the rush of sudden downpours.[19]

William Scully took an active hand in soil-conservation practices, and he had a major influence on the business routine established in the office of Scully and Koehnle in Lincoln. The partners there could expect a visit of three-months' duration from the landlord every two years. During the alternate year, one of the Lincoln partners would travel to Scully's overseas residence and spend three or more months there.

When Scully came on his inspection tours, he would carefully examine the books and visually examine the land. He had questions to ask about virtually every lease, and while he was not acquainted with each tenant, he became conversant with that tenant's ability as a farmer. In the early years, William Scully hired a fine team and buggy from the Lincoln livery stable and traveled through his estates. In Logan County he occasionally would stop, and very rarely he would visit with or eat a meal with a tenant who had been on the land for many years; but elsewhere he was quite aloof from his tenants. In later years, Frederick C. W. Koehnle kept a special matched team and carriage and a hired Negro servant to drive the landlord through his estates.[20] Agents in Nebraska wnd Kansas also supplied special vehicles for the landlord and went with him on his biennial inspection visits. On one such visit to Marion County, the agent, knowing of Scully's Irish background, thought he would enjoy meeting a typical Irish settler, so they stopped at the farm of Tommy Meehan.

"Meehan, meet Lord Scully," the agent said.

Meehan asked, "What did you say?"

Again the agent said, "Meet Lord Scully."

Meehan growled, "Get the hell off my land. He isn't anymore a lord over here than anyone."[21]

During each year that the landlord did not come to the United States, an agent would take copies of the rent rolls and certain other records for an extended visit with William Scully. One such visit by John Scully came in 1882, two years after John had married a member of a prominent Lincoln, Illinois, family, Louise ("Lily") Chamberlin. The young couple traveled from New York on a "vessel of the Cunard line. Mr. Scully goes on a business trip and will be gone three months. Mrs. Scully expects to spend most of the time in Scotland."[22] The return of the John Scullys in September was noted in the Lincoln newspaper. Koehnle, Scully's chief agent in later years, regularly made trips to Europe on estate business, although at times other agents from Lincoln were to serve as emissary to the landlord.[23]

John Scully developed into a leading businessman in Lincoln, Illinois, where he bought some Logan County land in his own name and was active in community business and social affairs. On June 16, 1880, he was married to Louise Chamberlin, who bore him two children, John Chase and Louise. On March 28, 1885, John returned from a business trip to Toronto, Canada, suffered an attack of erysipelas—a high fever caused by a streptococcus infection—and died on Saturday, April 4, 1885. William Scully, who was coming for an annual visit to the United States at the time, arrived in New York the following Monday. The funeral was initially postponed until Wednesday, April 8, so that William could get there, but travel time was too short. The funeral in the family residence in Lincoln was conducted by one of the young widow's cousins, who was from Chicago. The pallbearers were two members of the firm of Scully and Koehnle and four large landholders in the Lincoln area. Newspaper comments were laudatory of the integrity and character of the thirty-five-year-old John Scully. When his will was filed, the newspaper reported that his estate was valued at about $50,000.[24]

William Scully's administrative staff for his American lands had grown slightly during the years that John Scully was his chief agent. A full-time agent, Henry Fox, was employed in Grundy and Livingston counties, but elsewhere, agents worked part time for Scully. Alex Case, as part-time agent for Marion and Dickinson counties, could not leave his other work to go full time for Scully, so F. W. Fox, son of the Grundy County agent, was sent to establish an office in Marion. Other agents were dispatched to Beatrice and Nelson, Nebraska, where Scully's holdings were growing in value. The most significant changes, however, were the employment of Springfield resident Frederick Trapp and the renaming of the Lincoln-based Scully office to Koehnle and Trapp.[25]

Administratively, Trapp inaugurated many changes, including a record-keeping procedure for the Scully estates. Rental income in the years following Trapp's entry into Scully's employ shows a sensational growth. Records became more voluminous, and correspondence between the landlord and his agents in the Lincoln office was more complete. A later agent described Trapp as the legal and business organizer, saying that "when Fred Trapp came, the records started and the business really began to move. . . . he contributed much that could be of use to William Scully."[26]

After John Scully's death, land purchases declined in counties where William Scully already had land. Somewhat later, in the 1890s, extensive farm property was acquired in Bates County, Missouri; but more about that later. Land purchased in Illinois in 1889, 1892, and 1901 amounted to a mere 81.42 acres. Purchases in Kansas and Nebraska ceased in 1887. Anxious buyers made many attempts to acquire land owned by Scully, but he was no longer interested in selling, as he had been in the mid 1850s. He did sell railroad and highway right of ways and small tracts for schools or churches, usually stipulating in the deed that the land would revert to Scully or his heirs when it was no longer used for the stated purpose.

On one occasion in Marion County, Albert Crane, the owner of the very large Durham Park Ranch, wanted to trade half-sections with Scully, believing that it would be more convenient to both parties, as these tracts were isolated from other lands that they owned. If they traded, they would then have land in a block, for each had other holdings nearby. But Scully refused and suggested that Crane sell him the land in question. Crane, who was not anxious to sell either, had merely thought that the deal would be useful to both of them. Scully said: "It is a very serious thing to part with title to real estate. I will not do it."[27] Similar stories have circulated about his refusal to sell land; he was most emphatic about it. One of Scully's longtime employees recalled: "Mr. Scully did not like to part with land, once he had acquired it. . . . It is not a written rule, but an unfalteringly observed policy that Scully land will not be sold. Of course, Mr. Scully gave up land for townsites, school grounds, railroad rights of way and the like, but never disposed of any actual farm holdings except in the Oglesby case." That major exception was his sale in 1888 of 152 acres in Sangamon County, at $50 per acre, to Richard Oglesby, governor of Illinois.[28]

Administrative costs of the American Scully estates were estimated in 1890 to be $63,500 annually for the next three to five years. The 1889 figures had totaled $68,000, of which taxes represented $33,000. Other costs that year were for salaries of personnel in the Lincoln office—Koehnle, $6,000; Trapp, $2,500; Sims, $1,200; and clerks, $1,300; for Henry Fox in Grundy and Livingston counties, $2,000 each; for F. W. Fox, John Powers, and a man named Hazlett in Marion, Dickinson, and Butler counties, $2,500 each; for Leslie Gillett in Gage and Marshall counties, $3,000; and for Henry Fox, Jr., in Nuckolls County, $2,000. Other expenses reached $14,500, which primarily represented drainage development in Illinois; but earlier there had been a $200 road tax to be paid in Kansas, which was worked out by hired employees. Such activities, different in different areas, helped to justify the use of full-time agents in each state.[29] In 1891 the administrative costs showed only a slight change from the amount projected, and there were no changes for salaried employees.[30]

William Scully's excess capital in Scully and Koehnle's day was heavily invested in land, but considerable amounts were also put into United States governmental bonds and into bonds issued by the city of Chicago, the Illinois counties of Sangamon and Cook, and Russian railroads. Typically, these securities were purchased for him by his agents. As the Scully estates became more visible and their worth more fully recognized, the cost of amassing this huge estate seemed fabulous to onlookers. So, rumors circulated that the early-day purchases by William Scully had been financed by the European capitalist Rothschild. Perhaps, someone had seen a letter mailed to Scully's Lincoln office that had a Rothschild return address. The only connection between William Scully and the Rothschilds was that the interest on some of Scully's bond purchases was paid at the office of N. M. Rothschild & Son in London. During John Scully's last year as head of the Lincoln office, bonds from Hungary became a new attraction for heavy investments. Somewhat later, Scully purchased a large number of Uruguay Unified Bonds and bonds issued by either the city of Buenos Aires or the Republic of the Argentine. A last big purchase, in 1900 and 1901, was of Russian guaranteed gold bonds of the Wladikawkas Railroad.[31]

Apparently, Scully had always possessed a feeling of what were good securities to buy. Perhaps his experiences on the securities market alerted him to certain precautionary steps before buying. Eventually, he wrote these down and stated that the utmost quality "of an eligible Bond & stock [was] *Security*." His "Conditions of an eligible security" to be used as a guide by his chief agent, who would invest most of his surplus funds, were:

1st Don't deal in Faith . . . or in few ambiguous promises—(Deal in Fact—
2d Solid & well grounded bodies & power and will to pay out contract payments.
3d Integrity of the debtor—
4th A good percentage of interest.
5th Buy under par with as large a margin for appreciation as possible.
6th Cho[o]se a debt of large gross amount, so that more money can be invested in it without further troublesome investigations, & So that Bonds may be solidly known and dealt in, therefore easily salable & convertible—if necessary.
7 Long time (say 20 years, or over) to run, of Bonds . . . to avoid constant dangerous changes, & to preserve fair continuity—
8th No Bank, Railway, Commercial, or other trading, or venturesome or risky company (Speculation) is eligible
9th I prefer large Bonds of not less than $5,000. cash if obtainable
10 Coupons must be payable where they will not and cannot be taxed—
11 Bonds, repayable by drawings are very troublesome—
12 Don't lend to a Borrower, within about 35 degrees of the Equator.[32]

In 1901 William Scully stated that "it has been 32 years since we have had a lawsuit with a tenant over his lease or his rent."[33] An examination of the record shows quite a different story. Between 1870 and 1887 in Logan County alone there were forty-six cases with William Scully and two with his agents as plaintiff filed against tenants or former tenants for attachment, distress for rent, trover (to recover loss of wrongfully used property), assumpsit (an action to recover damages for a breech of contract), or replevin (recovery of goods and property).[34] According to Logan County records, most of these cases had a full hearing in court. No doubt the landlord was never present in any of these cases, and if they were important to him, he preferred to forget them in his old age.

In the 1880s there were two significant Illinois cases with William Scully as the defendant. The first was *The People* v. *William Scully*, filed in the Logan County circuit court in the January term of 1882. The action was initiated by state attorney Randolph B. Forrest to "recover back personal property taxes for the years of 1875, 1876, 1877, 1878, 1879, 1880 and 1881" from William Scully. The state argued that Scully owed personal property taxes of $200 to $350 per year on his rent rolls in Illinois, whereas the defense held that these rolls were owned in Great Britain, not in Illinois. The jury in this trial found for the state, and the Scully estates were ordered to pay $1,427.[35] Scully's attorneys appealed to the Illinois Supreme Court, which reversed the decision of the lower court.[36]

The other lawsuit involved the very large Hamilton tract in Sangamon County, which was purchased in 1877 for a total price of $215,297.40, the costliest single purchase that Scully ever made. At the time of the sale to Scully, the heirs of Archibald Hamilton, whose estate had obtained the land from the government in 1836, sold a land area that they believed to be 4,141.34 acres. The purchase, in a depression year, was $51.75 per acre, and Scully agreed that he would pay for additional acreage or get a refund for too many acres, provided that the property was properly surveyed within two years. The survey was not completed until 1885, and it showed that most of the quarter-sections contained about 163 acres and that the total land area was 71.03 acres more than Scully had paid for. Thus, on October 27, 1886, James C. and George H. Hamilton sued Scully for the amount of the extra acres plus damages, because he had refused to pay. In a court action that was appealed to the Illinois Supreme Court, Scully's prior payment was upheld. The survey had not been made within the prescribed two years, so the Hamiltons got no additional money.[37]

In the earliest days in Illinois, many of the Scully tenants were Irish or were from families of early American ancestry. But there were still tenants recorded in the earliest lease registers that had family names indicating German ancestry. However, the change in national origin for many Scully tenants coincides with the employment of Frederick C. W. Koehnle as one of William Scully's chief agents. German names crop up with increasing frequency in subsequent years, and later a report was made that most of Scully's tenants were Lutheran but did not attend church. In reminiscing about his estates in 1901, William Scully stated: "The majority of my tenants are Germans. As a class, I have found them the best farmers. There are dozens of German farmers on my lands who are rich men. There are scores of them who began as farm hands, then as lessees, and finally owners of farms."[38] Since many of these newer tenants were recent immigrants, it was believed that they would be less troublesome and more diligent as farmers than persons who had longer ties to the United States. Thus, Scully agents favored them.

Many tenants in Kansas and Nebraska also had German backgrounds, but the agents in their immediate area did not exercise a pro-German discrimination as strongly as did Koehnle. As a matter of fact, the location of Scully's land, near various foreign-language colonies, probably had as much influence on who were potential tenants as did anything else.

Even before he became a Scully agent, Frederick C. W. Koehnle had returned to Germany for extended visits. On one trip he left his family in the fatherland for a year and then returned to bring them home. Some accounts say that later he went to Germany to recruit potential settlers for Scully leases.[39]

By the late 1870s, when tenants with German names were replacing other tenants on Scully leases in Logan County, there were a few reports of bad feeling. One isolated incident was described by Lena Sparks, a daughter of one of the German tenants. She told of a rifle shot striking a post just above her father's head as he was sitting on his porch. Years later it was discovered that another Scully tenant had fired the shot because of his dislike for the new German tenants.[40] While the number of German names may have increased on Scully lease rosters, there were many other tenants with names of other foreign sources, as well as old-line American names. Scully's impression in 1901 may have been due to his conversations with Koehnle as much as anything else. Tenants were increasing in number by then, and there were many different names on his rent rolls. An examination of these lists does not fully support Scully's impression of where his tenants came from.

After Ballycohey, Scully devoted more of his time to his American estates, and his Irish holdings received far less of the landlord's attention. Perhaps due to the response of his peers to his actions at Ballycohey, Scully was motivated to seek a new permanent residence. Moreover, his estrangement with his family extended, in part, to his own three daughters. By using his growing wealth, Scully could live almost anywhere he wanted. After 1870, that desire was fulfilled by his relocation in England, without giving up his older residence at Ballinaclough. His move to England seemed to stimulate in Scully a desire to establish a landholding dynasty. Letters, brief memoranda, and occasional newspaper accounts provide the onlooker with a rare glimpse into the private life of a wealthy landlord.

wswswswswswswswswswswswswswswswswswswswswswswswswswswswswswswswswswswswswswswswswswswswswswswswswswswsws

7

The Private Life of a Wealthy Landlord

wswswswswswswswswswswswswswswswswswswswswswswswswswswswswswswswswswswswswswswswswswswswswswswswswswsws

By 1870, William Scully had become convinced that his permanent residence should be outside Ireland. Thinking it still unsafe to travel alone in his native land, he sought police protection everywhere he went. Frequently it was difficult to obtain police when he needed them. At this time or later he did not travel between his Irish estates without carrying personal firearms, usually a pocket revolver, and much of the time he had additional security from guards who went with him. During the year after the Ballycohey fight, there was public discussion in Tipperary about the additional police posted in four nearby barracks. Most of the debate was over the withdrawal of extra policemen without consulting the Tipperary magistrates. These magisterial officers, the real political power in the county, were incensed and critical of the removal of the police, and it became an issue that was taken personally by the commanding officer of the police forces who made the decision.[1] No doubt, Scully was grateful that the issue of police reinforcement, which he had brought on by his actions at Ballycohey, had passed him by to become a question for someone else to deal with.

There were few members of William Scully's immediate family left in Tipperary. Vincent, with only another year to live, still maintained a house at Mantlehill near Golden, but he made his home in Dublin or in London most of the time. Rodolph had lived in London since his marriage in 1850. Catherine, the sister who became a Spanish duchess, had died in 1867. The two youngest sisters were the only other siblings of William's who were still living. They were nuns on assignments on the Continent with the Society of the Sacred Heart.

Late in 1870, without giving up his house at Ballinaclough, William Scully moved to England, to make his residence first at no. 10 Dawson Place in the Bayswater section of London, a short distance north of the royal palace. In a manner typical of later years, he rented a fine home at a rate below what he believed it would have cost him to build, causing him later to repeatedly assert that "fools build houses for wise men to live in!" The only home he ever built was the white frame house in northwestern Logan County, Illinois, which was very large for the time and place and was expensive at $1,200 to $1,400. Even Ballinaclough House was not fully his possession, due to the complicated nature of Irish land law. In later years, in spite of his immense wealth, he would rent other homes and buy only one.

The fighting in the Franco-Prussian War had just concluded before Scully's arrival in London. Conditions in the English capital were in turmoil, even though the English were merely onlookers in this brief war which changed the map of Europe. In a letter to John Williams of Springfield, Illinois, Scully gave his impression, saying: "I arrived here safe—find things well. London pretty much crowded with Foreigners. Provisions of all kinds look to me to have a tendency to rise on account of the war."[2]

William Scully rigorously guarded his private life in the years after his removal to England, making it difficult to discover fully his character or even his whereabouts. He was described as taciturn, shy, and retiring. He had a lifelong reputation for being thrifty and close with his money but not miserly. Those who knew him well said that he worked hard and was an extremely capable and intelligent man. The nature of the activity of his landed estates placed ultimate responsibility on him for every change or any nonroutine function in his business. He fully accepted this responsibility and transacted all of his business through his agents, who were relatively well paid. Scully was regarded as generous with those close to him. Because there was no regularly imposed order of business in his daily life, Scully could take up matters at his own convenience and in his own time. By preference he did set regular times for doing certain things, and provision was made for reading, some of it directly related to his landed interests. Moreover, on his travels, Scully was observant of farming conditions and practices that were successful in different farming areas. Everywhere he went, at least in later years, he was described as bookish, because he usually carried an armload of books with him on his travels. About half of the books he read were related in some way to agricultural science and technology. The writings of Arthur Young and Robert Bakewell had influenced enormous changes in British agriculture, and Scully probably read everything they wrote.

Thus Scully gained an increasing knowledge of agricultural science and technology as his extensive landed estate grew in size. In spite of the general assumption that the rich fertility of the Illinois land would last forever, he became convinced that positive efforts were needed for protection of that land. He believed that the best long-range goal for his land was to carry out practices to conserve the soil's fertility. Thus, as he inspected his own properties, he carried little sacks of soil away from each farm. His tenants talked among themselves about their eccentric landlord who was making a collection of soils from each of his farms. Actually, he was getting samples from which he could learn the characteristics of his land through chemical analysis. Because of the information derived from soil samples, he incorporated changes in his leases. Thus came the requirements by the late nineteenth and early twentieth century, in all American Scully leases, that red or white clover—or alfalfa on certain parts of his estates—be grown on one-fifth or one-fourth of the cultivated acres. In addition, there were special provisions that, after eighteen months, these legumes be plowed under as green manure.

In all probability these requirements for the planting and use of legumes on his land were slow in coming for William Scully. Generally, he was comparatively early in developing these requirements in the United States, but he had owned his land a long time before they were put into effect.

Scully's next change of residence, following a brief return to Ballinaclough, seems to have been directly related to his search and drive for better agricultural techniques for his estates. That was his move to the village of Hatfield in early 1872. Hatfield in Hertfordshire, some fifteen miles north of London, was only a few miles from Harpenden and its famed Rothamsted agricultural experiment station.[3]

The old 600-acre Rothamsted manor was inherited by John Bennet Lawes in 1822. Soon after leaving Oxford in 1834 he began crop-rotation and fertilizing experiments on his farm, and in 1842 he discovered and patented a process for making a financially profitable fertilizer, superphosphate. Superphosphate was made by treating phosphate rock with sulfuric acid, and Lawes established a factory to produce the fertilizer in commercial quantities. The year 1843 is usually treated as the founding date of the agricultural experiment station, as Joseph Henry Gilbert, a chemist, came to work then and began a lifelong collaboration with Lawes in agricultural science. Rothamsted continues its work in agricultural experimentation more than a century later and is recognized as the oldest continuously operated agricultural experiment station in the world.

Lawes and Gilbert, two giants in extending the borders of English agricultural science, were near the peak of their productive activity when Scully moved to Hatfield. Lawes had been elected a fellow of the Royal Society in 1854, and in 1867 he and Gilbert were jointly honored with the Society's royal medal for their outstanding work. The use of superphosphate grew rapidly in these years and was of prime importance to certain kinds of crops in England and Ireland. The four-field Norfolk system of crop rotation, with one field each year in turnips fertilized by superphosphate, showed beneficial results and attracted much attention throughout the British Isles.

Scully made a direct application of the superphosphate experiments in the 1870s, when he purchased a large quantity of buffalo bones in the Great Plains area of the United States. The railroad "gave him a very cheap rate to Philadelphia and there he hired a sailing" ship to take the bones to Waterford, not far from Tipperary, where "he had the bones all smashed up and he put them on the Scully Estates in Tipperary."[4] Presumably, he had the powdered bones treated with sulfuric acid before they were spread on his land.

William Scully's next residential move was to 70 Holland Park, Kensington, county of London, where he had moved into an expensive house by February, 1874. His three daughters by his first wife had disappeared from his life by this time, although he continued to provide for them in certain financial ways. In Kensington, Scully organized his staff of servants to run his house in this new prestigious location less than a mile west of the royal palace. He became acquainted with his neighbors, one of whom, John Chynoweth, was of Welsh heritage. Chynoweth, his wife, and daughters lived at 35 Holland Park. Because Chynoweth and his wife had been married at Vera Cruz, Mexico, aboard the United States frigate *Cumberland* by the chaplain in April, 1848, it was frequently reported that Mrs. Chynoweth was of Spanish origin. Both of her parents, however, were Cornish, and her maiden name was Harriet Brobenshire. John Chynoweth's years in Mexico, as a banker, businessman, and owner of silver mines, spanned the period from Mexico's war with the United States through the era of the reign of the emperor Maximilian (1863-67). He prospered in his Mexican adventure, so he could easily afford the comfort and luxury of a big house in Holland Park.[5]

Scully played backgammon with Chynoweth on many long winter evenings, and he became acquainted with members of the Chynoweth family, particularly the attractive eldest daughter, Angela, who, like her two younger sisters, had been born in Mexico and was thoroughly familiar with Mexican culture. Enriqueta Angela Lascurain Chynoweth had been born on March 29, 1849, about a year after her parents' marriage. As a "spinster," a month before her twenty-seventh birthday, she was married to William Scully, who was almost precisely twice her age, on February 16, 1876. The marriage, "according to the Rites and Ceremonies of the Established

Church," was performed by the Reverend John Robbins, D.D., in St. George's Church, Campden Hill, county of Middlesex, with the bride's parents as official witnesses.[6]

For many years the Scullys made their London home at 70 Holland Park, then they moved down the street to no. 12. After the death of Angela's parents, her sisters continued to live in the home at 35 Holland Park, maintaining the cultured life of English gentlewomen who had their own carriages and horses.

Late in 1876, William John Chynoweth Scully was born to William and Angela Scully.[7] As their first-born male heir he had much to look forward to, and "Willie" became the pride of his parents. Thomas Augustus, their second son, was born on September 16, 1878, and was named for William's favorite brother. Slightly more than a year later, their daughter, Angela Ita Harriet, was born; and Frederick, their third son and last child, was born on October 2, 1881. Four children were born to William and Angela in slightly-less than five years, and then they had no more. A landlord dynasty seemed assured!

Angela organized the duties of her large staff of servants to be as efficient as possible. She wanted her household to run without seeming direction, and she spent much time in training her servants. In later years more than a dozen servants were needed to care adequately for her home, with additional ones to handle the horses and carriages and other outside work. Proper response to their duties would assure her servants of life-time positions, should they meet her expectations.

Mrs. Angela Scully had been well educated by governesses, and she easily fit into the pattern of a wealthy matron of culture in London. Her children received the best education that money could buy, and they attended prestigious schools. The children's needs were cared for by nurses and governesses who were always available to them. There was a certain aloofness in the parent-child relationship in this family. Much of the time her children addressed her as Mrs. Scully. Thomas's earliest memory was not of his mother or father, but of his German nurse.

In time, the family began to make regular trips to Cannes on the French Riviera, where they rented a villa each winter season. Preparation for these long trips, by rail, with a channel crossing in between, were remembered with horror by the children. Restrooms were either not available or extremely dirty along public transportation routes, so the children were given a powder to purge their systems before going on such a long trip. Even with these precautions, they had to visit every w.c. they passed, and the parents could not understand why. It was recalled in later years that on these visits to Cannes a nurse would take the children out to "watch Queen Victoria go by."[8]

Two of the children, probably Willie and Tommy, went with their parents to the United States in 1885, arriving in Lincoln, Illinois, a few days after the funeral of John Scully. Also in the traveling party was one of Angela's sisters, who was "constantly shocked by the boys antics" and their restlessness.[9] There were other times that Angela's sisters went to Cannes with them. The activity of the young Scully children was appalling to their maiden aunts, and the children objected strongly to the restraints imposed on them by their mother's younger sisters.

After a brief stay in Illinois in 1885, the Scully party, along with Mrs. Louise Scully (John's widow) and her two children, departed in early July for St. Louis, where they boarded a Mississippi River steamer for an upriver journey to St. Paul and Minneapolis. There the children were left, and a three-week trip was taken by rail to Yellowstone Park, which had been established as "a pleasuring ground" by Congress in 1872. Fort Yellowstone, which was maintained by the army, guarded the few visitors to Yellowstone's splendors in the 1880s. Only a few years had elapsed since Chief Joseph and his Nez Perce, in hostile array, had crossed the park. Mrs. Louise Scully returned to Minneapolis by the end of July and stayed there another month, while William Scully, with his wife and sister-in-law, went on west to Seaside, Oregon, where they spent a month in leisure before returning to Lincoln in September.[10]

On one of his trips to the United States, possibly in the mid 1890s, William Scully was accompanied by Angela on a visit to the western lands. They arrived in Marion, Kansas, at a time when drought, hot winds, and grasshoppers were causing extreme shortages. They stopped at the best hotel in town, where William Scully sought out the manager of the dining room after one meal and said: "I want to offer you an apology. I noticed today when we left the table that Mrs. Scully left a small piece of bread on her plate. I assure you that this will not happen again."[11]

William Scully had a much closer relationship with his sons than he did with his daughters, but even with his sons a clash of strong personalities was evident. The vast differences in age between father and children may have accentuated disagreements, but the only daughter of his second marriage recalled that she never got "along with her father" and that he never discussed his business affairs with her. Scully's plans for his sons included instruction in management of his estates and an understanding of their future role as landlords. They were sent to outstanding public schools and then to colleges such as the Royal Agricultural College at Harrow and to Cambridge University. About the only time the sons were with their parents was during vacations from school. Eventually the sons came to recognize the immense foresight and great patience that William Scully had exercized in accumulating his great landed estate, and they came to realize that by all means they must "hold on to the land."[12] Since land was the basis for their immense wealth, Scully impressed on his sons the "necessity of avoiding encumbrances on the land." Land to him was a permanent investment, not mere speculation. Scully's estate managers have always pointed with pride and some astonishment to the fact that there were no mortgages on any of the Scully holdings and to the relatively few entries on all Scully land abstracts of title. Part of the sons' education came when they traveled through the Scully estates with their father. They would dig holes in the land with a post-hole auger, and they would get instructions on the various kinds of soil and subsoil on their land.[13]

William Scully's attention to detail for the improvement of the Scully estates, especially in America, was considered phenomenal. He combined a visionary dream of a great landed estate with practical considerations such as the employment of able, scientifically trained agriculturalists to perform certain management responsibilities on his estates. Several of his American agents were sent to neighboring universities for specialized training, just as he had sent Darby Gorman, his Irish steward, to Edinburgh University.

One of Scully's peculiarities in conducting his business was "never to permit a blotting pad to be put upon his signature after he [had] written it." He would simply say, "Let it dry, there is no hurry."[14] Much of his business was conducted with grave deliberation, quite in contrast to the impetuous behavior he had demonstrated at Gurtnagap and Ballycohey. When he was severely criticized in lengthy editorials and in libelous newspaper comment in America, Scully reacted by remaining mute. "Not one word of defense ever came from him. He did not talk. He did not write. . . . His policy was silence. He went straight ahead with his work."[15]

Part of William Scully's personal philosophy can be seen in the brief memoranda he wrote in the 1880s and 1890s. Some of these remarks were based on widespread reading and reflection and were of an extremely conservative nature. For instance, on June 5, 1890, he wrote: "The following seems to be the cause of the ruin of States," and he listed three brief points:

> 1st Subversive & demoralizing doctrines—
> 2d A feeble administration—
> 3d The liability of mankind to periods of semi-insanity—
> The above causes united, create a violent ferment—
> Then, if the body politic retains abundant internal vitality, it cures itself (as in a fever,) by revolution and Reform—If it has not this vitality then the State dis[s]olves—[16]

Based on his own observation and from reading, Scully believed that there was much degeneration in the period between 1873 and 1881. He wrote that

> the British Government has proven radically and hopelessly bad. It is now a mere democracy, almost uncontrolled by the Crown or by the house of Lords. The scum rises to the surface. There is [incipient] change and tinkering with the laws, especially with the land laws, which above all others should be stable, because [of] dealing with a property that [possesses a] title.
>
> The condition of things in both England and Ireland may be expected rather to be worse than better. It will [go from] bad to worse, until some terrible catastrophe overtakes the British Government & people, and until disaster & misfortune . . . restores [them] to sanity & to better habits & morality.
>
> Most in Ireland and many in England of the richer & upper classes are indolent & luxurious. The lower and increasing classes are selfish and disloyal, a prey to agitators. In Ireland they can only be restrained by armed force. This force no longer exists because that poor [classes] are now the governing classes, & will not use that force effectively against themselves.[17]

As their children were growing up, William and Angela Scully were described as persons of great dignity. William wore "dark clothes, with an old-fashioned black bow tie." Out of doors he wore "a skull cap and black shoulder cape." He was tall and slim and, as he aged, partly bald and "slightly stooped." His features "were sharp and intelligent," and "his blue eyes, while not bright," were penetrating to visitors. An observer in later years described him as a "careful, interesting talker" with a "slight impediment in his speech . . . not noticeable when he talks rapidly."[18] One Missourian, after seeing Scully riding "in a surrey with a plain shawl fastened across his chest," said that "he looked like Gen. Grant."[19]

Angela Scully was a typical nineteenth-century *grande dame* who became a rather large woman in later years. On one of her trips to America she arrived with a maid and seventeen trunks. She was frequently seen wearing "a black silk old fashioned dress probably not changed in style" for many years. She was interested in "running a perfect household, always paying her bills on time, and being kind and severe to those who worked for her or lived on the Estate." Her attitude toward servants and those around her reminded some who knew her of what they had heard about Queen Victoria, and it made them wonder if Mrs. Scully were modeling her behavior after that of the dowager queen.[20]

Those in close contact with William Scully vouched for his "fine character," and they claimed that they could "trust him implicitly." His own explanation for the great size of his estates was that he had concentrated on a particular goal and that he had worked hard to achieve that objective. His direct answer to a question about why he was successful in an agricultural enterprise when so many persons failed was, "As some one man must outstrip his fellows in each of the works of life, so will I excell as a farmer, making it the goal of my life that I shall become the first farmer in the United States."[21]

The deliberate nature of Scully's business activity in the United States was reflected in his dinner each evening. He was "never known to haste." Ordinarily it took two hours to eat an evening meal, and if there were a formal occasion, he would generally "make a night of it." A newspaper reporter said that William Scully neither consumed alcohol nor used tobacco.[22] Other stories disclose preparations for Scully's visits to each of his agencies which required the purchase of fine Scotch whiskey. In his office in Lincoln, each afternoon about four o'clock he would put an inch of whiskey in a glass and go to the water fountain to fill it up. Then he would drink sparingly. While not a regular attendant, he was a member of the Church of England. There are no reports of his being a member of any of London's many clubs—or of his attending any theatrical or musical events.

In Ireland, the primary criticism and heated newspaper comment directed against William Scully came in the 1860s, during the trials involving Gurtnagap and during the affray and aftermath of Ballycohey. His actions on his Kilkenny and Tipperary estates had gained notoriety for him in the British Isles. He sold Ballycohey in 1868, but he still possessed almost three thousand acres of land in two Irish counties in 1875, and he never disposed of these holdings, which were the accumulation of many generations of Scullys.[23]

It was more than a decade after Ballycohey when Scully became the object of vigorous criticism in the United States. The agrarian upheaval of the Granger movement of the 1870s had given way to the more active and aggressive Farmers Alliances of the 1880s. Then a number of newspapers launched an attack on large-scale alien absentee landownership in the United States, with the object of driving such landholders from the country. Some of this effort concentrated on Scully because he fit the picture so completely of the absentee alien landholder and because his estates were so huge. Ironically, he had been called an alien in Ireland, and the pro-Irish press in Chicago and a number of other places resurrected the anti-Scully biases from Irish incidents of the 1860s. Most of the anti-Scully newspaper activity was from communities that were quite near to large Scully holdings. As one newspaper headlined a story, they were "Skinning Skully"; and that pattern produced a new rash of Scully stories, many of which were erroneous and far-fetched. However, these accounts promoted a popular concept about the Scully lands that has become entrenched in historical memory throughout the country.

wswswswswswswswswswswswswswswswswswswswswswswswswswswswswswswswswswswswswswswswswswswswswswswswswswswswsws

8

Scully's Scalpers

wswswswswswswswswswswswswswswswswswswswswswswswswswswswswswswswswswswswswswswswswswswswswswswswswswsws

Thomas Jefferson's highest ideal for the republic of the United States of America was a nation of citizen-farmers who tended their own land. Tenant farmers, to him, were of a much lower order than the stalwart farmer-yeomen, who were the backbone of the country. To Jefferson, tenant farmers lacked the independence needed to make the nation strong; they were of little more value than clerks and workers in a factory or mine who could be exploited by their employers. In Jefferson's day the vast American territory lay waiting for their owner-farmers who would possess a holding, more than adequate for that time, of eighty to one hundred and sixty acres.

But not everyone who wanted a farm in the latter half of the nineteenth century could get one, even though widespread land ownership became a reality. The Jeffersonian view continued to be strong. Pressure for "free land" and the disposal of government tracts in areas as small as forty acres did lead to easier acquisition by potential farmers with little capital. The Homestead Act of 1862, the cherished product of a generation-long struggle for free land, offering a quarter-section of government land, was an extension of the Jeffersonian dream for an egalitarian America. Other land laws, such as the Pre-emption Act of 1841 and the Timber Culture Act of 1873, promoted the "superior" rights of the settler and occupier of the land, as compared to the speculative land seeker who hoped to gain from resale at higher prices. The large-scale landowner in the "Jeffersonian dream" was called a land monopolist, and monopolies of almost every kind and description were abhored, condemned, and restricted in the America of the late nineteenth century.

It was in this era that the members of the Patrons of Husbandry, the Grangers, advocated restraint on railroad monopoly in Illinois and elsewhere, and they gained legislation restrictive on railroads in the 1870s. As sons of the soil, they advocated easier entry into farm ownership. They were quick to criticize large landowners for holding on to land that, for the good of the nation, should be in small owner-operated farms. As a landowners' movement, Grangerism failed to consider tenant problems. However, as the Patrons of Husbandry declined in political significance in the late 1870s, the anti-large-landholder views continued strong. In addition, people in rural areas looked upon the absentee landholders as noncontributors to the development of their region who would gain in the overall increase in land values without providing input that was commensurate with their share of ownership. In the post–Civil War era a surge of patriotic fervor also castigated the "alien" property owner in the United States whose ownership denied the full use of certain lands to citizens. The realization that land in Illinois and other western states was already in private ownership, or soon would be, brought on a public outcry against owners of vast territories. The ultimate object of criticism and complaint was the alien absentee large-scale landholder, personified by such landlords as William Scully.

Probably the first published criticism of Scully's cash-rent policies in America was printed in Chicago and Pontiac, Illinois, newspapers late in 1878. Also attacked was Scully's relatively new policy of guaranteeing rent payment by not allowing a "tenant to sell any of his crop of this year until the rent has been paid in full."[1] Later a Chicago newspaper asserted: "[Scully] gets his share of bad tenants. Some are shiftless and waste their time and money in drink. Some are industrious enough, but try to beat their landlord out of all they possibly can. Such are usually behind in their rent and stringent measures are sometimes taken to compel them to pay up. Now and then a tenant will haul off his grain in the night, succeed in selling it, and 'skip-out' without paying his rent."[2]

Tenants held protest meetings, because of the new restraint on selling their crops, in order to seek intercession from the governor. The Chicago *Times* held that "Scully's actions savor much of the tyranny of the absentee landlords of Ireland."[3] These early attacks on Scully drew little public response, but they set a pattern for later protests. The lack of public clamor in 1878 can be seen in the passage of an Illinois law that ratified the very rent requirements that Scully was already imposing on his tenants. Provision was made that enabled landlords, when they expected tenants could not pay their rents, to "institute proceedings for distress" prior to the date when rent was due.[4]

As might be predicted, the New York *Irish World* had something to say about Irish landlords who transferred their assets and abilities to the United States. In an 1880 discussion of the problem in fixing cash, as compared to share rent, Scully was mentioned in passing, and landlords in general, particularly the ones who imposed cash rent, were attacked:

> Mark the coolness and impudence of this fellow. He starts out on the evident assumption that the lord of the soil is the natural keeper and protector of the servile farmer. He affirms that at present many tenant farmers are coming to this country from Great Britain who would like to rent farms for a long term of years. He is right but tenant rack-renters are coming too. A Marquis of Sligo and Lord Oranmore finding it too hot in Ireland can come over and set up their traps on these shores, as Scullys and Sullivants. The poor, hunted rat of a peasant has simply jumped from the hopper into the flame, and will he so far forget himself as to say that he left his interest in the Land Question at his battered door in Ireland.[5]

Also that year a local correspondent for the Lincoln (Illinois) *Times* placed in his column an attack on Scully that proved unusually embarrassing to the newspaper. He said:

> Broadwell
>
> "Lord" Scully, the monstrous land tyrant, and curse of Logan county, is justly entitled to the supreme contempt of every man having within him a spark of honesty. Last fall this scoundrel promised, through his representatives, to give tenants in this vicinity an abatement in their rents. Promises were also made to pay for trees long since set out; but now the fulfillment of both promises is refused.
>
> Gullivar[6]

Jumping on that story, the Chicago *Times* reported that "considerable talk is credited in this vicinity by the announcement that Mr. Scully, the noted Irish landlord, threatens to bring suit against Mr. Stokes, of The [Lincoln] Times . . . for $5,000, for defamation of character," and the Chicago paper reprinted the Broadwell account.[7] Apparently the retraction, published the following week in the Lincoln *Times,* satisfied the Scully agents, as there was no resulting lawsuit. The *Times* stated under an "EXPLANATORY" headline, which was longer than the original story causing the trouble, that

> an apology is due for the publication of an item in our Broadwell correspondence which, by an oversight of the editor, appeared in last week's issue of the Times, reflecting upon the personal character of Mr. Wm. Scully and the business integrity of his agents, Mr. John Scully and Fr. C. W. Koehnle, and we regret that anything should have appeared in this paper assailing the private character of these gentlemen. Their many friends and acquaintances in this county bear testimony to their integrity and upright characters as business men. The Times has heretofore attacked and will continue to attack a system which permits any person to acquire so large a body of land as is owned by Mr. Scully in Logan county, but will not knowingly lend itself to the . . . abuse of any individual. The aim of the Times is to control false principles and false systems, not to attack the private character of our citizens.[8]

The delineation of how far they would go in criticizing the Irish landlord may have prompted correspondent Katie to write the following in the "Prairie Creek" column of the Lincoln *Times* several months later:

> I wish for the good of Logan county to call attention of your readers to an idea of mine in regard to Logan county's curse. I mean William Scully. I have talked with some of my neighbors and we think we ought to have an act passed by our legislature imposing a special nonresident tax on all lands in the state owned by men born and living in foreign countries and to such an extent as to compel them to either sell or come to this country to live.[9]

Either Katie struck a responsive chord in Logan County, or she was voicing a common idea. A short time later the state of Illinois sued Scully for back personal-property taxes, as mentioned previously. The decision in the local court found for the state, but was reversed on appeal to the state supreme court. During this court case a neighboring newspaper, the *Republican,* printed the most scurrilous and libelous of the attacks on Scully up to that time:

> "Lord" Scully is an Irish leech who can scarcely write his own name. His wealth, obtained by grinding the life blood out of tenants, gives him his title. Scully has been one of the worst curses that ever afflicted Illinois. He is so unanimously hated that he dare not ride over his own possessions in Logan county for fear of assassination. He shows no mercy to his tenants and even forbids any merchant to grant them credit until they show a receipt that their rent is paid. He never lets them get a start in life, and prevents the county from developing. Every short time he adds [a]

> few hundred acres to his domain, and his continued encroachments excite no little apprehension. He is an old man now, and some of these days he will die. This is the only consolation in sight.[10]

Thus, "Scully's Scalpers" developed a popular theme, and newspapers that were slow to join the attack were frequently identified as Scully organs.[11] But the opening volley against landlord Scully in the years through 1883 was as nothing to the barrage of anti-Scully editorials and stories that came from the prairie counties in Illinois, Kansas, and Nebraska in the years 1886, 1887, and 1888. The newspaper editors in Kansas and Nebraska, especially, promoted an unrestrained and vituperative agitation that might have had some impact on the landlord and his plans.

The general theme of these attacks on Scully was to condemn alien absentee landownership and to demand legislation making future land acquisitions by aliens illegal, with no right of passing that property on to heirs. During the late 1880s, Scully's agents actively bought land, and these purchases were highly publicized. The purchase of the 1,500-acre Cayuga tract in Livingston County, Illinois, for instance, was ill timed, as it came in 1887, at the height of the campaign against Scully. Local comment labeled it "very unfortunate" and generally injurious to the community.[12] Ironically, it was the poorest land that Scully owned in Illinois, and stories were told that he thought he was getting land six miles farther south, where the soil was superb.[13] Because Scully's purchases in Illinois, Kansas, and Nebraska almost completely ceased after 1888, it seems likely that the landlord decided to avoid the hostility that he engendered when he bought additional land in the counties where he was already established.

Key issues in the attack during the late eighties on Scully were: (1) his alien status; (2) his absenteeism from his lands; (3) his cash renting system; (4) his requirement that tenants pay taxes and other land assessments; (5) his one-year lease; (6) his refusal to provide improvements on his lands; (7) his refusal to sell any of his land; (8) the "burdensome requirements" that became an integral part of his formal leases; (9) his lack of acquaintance with tenants, and his unwillingness to communicate directly with them; (10) the quality of the tenants that were attracted to Scully land; and (11) his lack of public relations to explain why he was developing certain programs for his lands.

Most newspaper accounts that were critical of Scully and his role as landlord mentioned his alien absentee ownership, and they publicized efforts to find lawful limitations on alien absentee ownership of land. This campaign was associated with demands that unused railroad land grants be returned to the government and that public lands be closed to large-scale purchasers. All national political parties joined this fight as early as the election of 1884. Eventually, ten state legislatures and the United States Congress passed legislation that was directed at landlords such as Scully, and in many cases he was mentioned by name. Indiana passed restrictive legislation in 1885, and two years later, Indiana was followed by Wisconsin, Minnesota, Colorado, Nebraska, and Illinois. The antialien bill in Nebraska was introduced by Relzy M. Aiken of Nelson, the representative from Nuckolls County. The legislature quickly approved the measure, which prohibited purchase of land by nonresident aliens.[14]

The Illinois legislation was introduced by Representatives Pierce and Kreitzinger, both of Logan County; John Piatt of Henry County; and Virgil Ruby of Piatt County.[15] Pierce was a farmer on land adjoining Scully's estate who had once rented eighty acres from Scully. The Chicago *Morning News* reported that his "antagonism to Mr. Scully is said to date from the time when he was a candidate for re-election as supervisor, and Mr. Scully's tenants flocked up in large numbers to the polls and voted against him, thereby securing his defeat."[16] Generally, however, the Illinois antialien measure was identified as "Mr. Ruby's bill," and it had little trouble in becoming law. However, before it was approved on June 16, 1887, it was divided into two laws, the first prohibiting "non-resident aliens from acquiring" land in Illinois and the second preventing "alien landlords from requiring tenants to pay taxes assessed upon the land they rented."[17] The first law was tested in the courts and was appealed to the state supreme court, where it was found to be unconstitutional, necessitating a more refined law in 1897. In response to the second law of 1887, Scully merely altered his leases so as to include an added amount of cash rent which was equivalent to the taxes that were assessed against the land.

Legislation against nonresident alien landownership was passed by Iowa in 1888, followed by Idaho and Kansas in 1891, and Missouri in 1895. In Kansas, Senator Alfred Lee Redden of Butler County introduced Senate bill no. 1 in 1887 to "prevent the acquisition of real estate by aliens and foreign corporations." Because such legislation violated a portion of the Bill of Rights of the Kansas Constitution, the move was quickly directed to change the constitution, with an amendment originated by Senator R. C. Crane of Marion County.[18] Strenuous efforts by editor Whitaker of the Marion *Register* made it the leading agitator for restraints on Scully in Kansas. The *Register,* when less than four months old, had its first story on "Tyrant Scully" and the action of the congressional judiciary committee that had reported adversely on national legislation.[19] Through the first half of 1887 almost every issue of the *Register* contained an anti-Scully editorial or inflammatory story. Scully was referred to as the "Bum," and predictions were made that "his bumness" would have to pay taxes on railroad bond voted in the county if farmers "would absolutely refuse to rent his land." Whitaker urged an "unrelenting boycott . . . which would force Robber Scully to sell his land," and he added: "Keep up the agitation; it accomplishes wonders." Reports abounded in this paper of favorable response in other newspapers to the campaign of the *Register* against "His Royal Nibs." The nearby Peabody *Graphic* portrayed the Marion paper's action as "Skinning Skully."[20]

In February the *Register* sent petitions throughout Marion County to get support for legislation pending in the state legislature. Two hundred copies of a special "antialien bill" issue were distributed to legislators. Also that month a letter was printed from an Ohio writer, urging every "citizen of Marion county [to] enlist in the fight and do all in their power to rid the county of that tyrannical landlordism which is keeping many away, and pauperizing the few who have ventured in on the land."

The rival Marion *Record*, whose editor had complained in 1884 that Scully owned "more land in the county than any man ought to own anywhere," actually entered the campaign against Scully somewhat later.[21] When the *Register* was heating up on Scully, the *Record* published a letter from a "Scully Renter" who described his experiences of the past four years, in three of which he had paid half the income of the land to another landlord and the past year, when he had "just as good land of Scully for $2 to $3 per acre and not bothered by landlord running 100 head of cattle in the field right up to corn planting time." This positive Scully report was countered by the *Register* with a list of persons who thought Scully was "a curse to Marion county farmers." While the Scully renter said that his "lease binds man to what he should do," and it takes "a rustler to make a go of it on any land in a dry season and Scully land is no exception," the *Register* reported that additional newspapers over the state were taking up the campaign. "Anti-Scully" continued to be its war cry as the *Register* resurrected terms such as "feudal system as it prevails in Ireland" and asserted that "Scully has done his best to transplant his system of rackrenting in this commonwealth."[22]

The Kansas election on the antialien constitutional amendment came in 1888, after the *Register* had ceased to exist. With every county approving the new amendment, it received an overwhelming 220,419 yes vote to only 16,611 against it. In Marion County the vote was 3,178 to 134. Editor E. W. Hoch, of the Marion *Record*, belatedly supported restriction on alien landownership in Kansas, and as a member of the house of representatives in 1889, he introduced a bill to implement the new constitutional amendment. Similar legislation was introduced in the senate, but the two houses failed to agree on a new law. Not until 1891 did the legislature respond to the message of Governor Lyman Humphrey requesting a law to implement the constitutional amendment. This bill was introduced and was strongly supported by senators from Saline and Morris counties, who were well acquainted with the Scully "menace." The governor identified the required law as one that has the "well-understood purpose" of "prohibiting the accumulation of vast landed estates by aliens."[23]

Federal legislation "to restrict the ownership of real estate in the Territories to American citizens" was approved on May 3, 1887. Like other antialien legislation in Illinois, Nebraska, and Kansas, it was introduced by a member who had personal contact with Scully's estates. This act was brought forward on July 31, 1886, by Congressman Lewis E. Payson of Pontiac, Illinois. When the bill came before the committee of the whole, Payson said that "in the hands of these foreign owners and holders these lands are made subject to a system of landlordism and conditions totally un-American, and kindred to that existing in the Old World, systems and conditions that have spread ruin and misery wherever they have existed in Europe." When the antialien bill came to a vote, the House supported it 210 to 6, with 106 absent or not voting, and it had a similar easy path in the Senate.[24]

Even national journals, at a great distance from the Scully lands, carried accounts of the actions of the landlord's estates, at times taken from newspapers situated near Scully's operations. In 1886, in an article entitled "An Oppressor of His Tenants," the New York *Times* reported that Scully was "one of the chief figures among the alien proprietors of American soil, and has introduced the meanest features of the worst forms of Irish landlordism on his estates." It further held that in Logan County, Illinois, alone, Scully had "reduced 250 tenants and their families to a condition approaching serfdom." A little later a report on a tour of Nebraska and Kansas found "between 60 and 70 families [who] have bound themselves to Scully, and so long as they remain on his land they cannot throw off the yoke that holds them in slavery." The lease, which bound the tenant to Scully, was "virtually a chattel mortgage," and before signing it, the renter had to "waive all rights of exemption."[25]

In 1888 the New York *World* reported exuberantly: "Landlord Scully Selling Out"; and a little later the Philadelphia *Evening Star* headlined a story: "An Evicted Lord. Mr. Scully Driven from Illinois." This account reported:

> The evictor has become the evicted, and the man who has driven tenants from their holdings is deserting his own possessions. William Scully alias "Lord" Scully, the alien landowner against whose foreign methods of landlordism the Alien Land law adopted by the last Illinois Assembly was directed, has resolved, rather than abandon his Tory methods, to get rid of the Illinois

> lands. The extent and fertility of these lands and the Scully mode of collecting rents and paying taxes were fully described in the Herald last March.
>
> At that time the Alien Land law adopted by the preceding Illinois Assembly was still largely experimental in the application to the evil for which a remedy was sought—the reversion of the property to the State, in event of "Lord" Scully's heirs or assigns failing to become citizens . . . . Public opinion was aroused not only in Logan and Livingston counties, where the great body of the Scully lands are situated but all over the State, and the result is that Scully will sell out.[26]

It is unlikely that the press consulted either Scully or his agents: these stories were largely wishful thinking. But unknown to the press, William Scully was seriously considering the sale of his land. In a manner that he had used before, he wrote a memoranda in July, 1888, which he entitled "To be attended to—In Sales of American lands by me."[27] The following points were recorded, in case a decision was made to sell:

1st	sell	subject to Existing leases rights of way of Railroads & rights of others
2d	Do	at a lump sum & not at an acrable [an amount per acre]
3d	"	at either a Quit Claim or at most a special warranty Deed only—warranting only against acts done by myself—
4th	"	All purchase moneys to be paid only to my authorized Bankers—Perhaps (if I find fit) Interest may be collected by agents—
5th	"	My wife and I to convey lands sold in Ills., "according to our respective Estates & Interests therein."
6th	"	Specify clearly who pays taxes—for the past & future years—
7th	"	Specify when purchase money should be paid—say by March 1 to bear 6% afterwards—
8th	"	Coal underlying can be strictly reserved (if wise to do so)—

Obviously, William Scully came to the conclusion that he could survive with the new Illinois law, and his decision against selling any of his many farms was stronger than ever. The threats of forfeiture of land owned by nonresident aliens was never considered seriously in the Illinois Assembly or in other state legislatures. The most that could be expected by these actions, which responded to a popular mood of citizens in each state, was that Scully could not buy additional land unless he became an American citizen.[28]

Another action taken by Scully at the time of his memorandum defied legal explanation. In London, on July 10, 1888, he deeded all the Marion County land to E. Angela Scully, his wife. Perhaps similar deeds were recorded in each of the other counties where he had land. A second deed was made out the same day, deeding all the Marion County land back to William Scully. Both deeds were signed before Thomas M. Waller, consul general of the United States in London. Both deeds were filed with the Marion County register of deeds on September 26, 1892, the first at ten o'clock A.M. and the second, forty-five minutes later. Perhaps Scully was apprehensive of a situation in which he would want to quickly shift ownership. By filing the right deed, the title would appear to fit the situation. The deeds, as they were filed, did not change the ownership of the land under Kansas law. It was the same as if they had not been executed.[29]

Governor Richard J. Oglesby, who purchased a quarter-section from Scully in 1888, was interviewed by a reporter for the Chicago *Tribune* early in 1886 about land laws, public lands, tenantry, and land tenure in England, Ireland, and the United States. In writing to Scully about the interview, the resulting story, and Scully's response, the governor said that the reporter

> took occasion to make many allusions to you personally, and to your history; that after talking some time and obtaining expressions from me concerning our public lands, and in reference to you personally, (in which I disabused him of what I supposed to be certain erroneous opinions about you and your system of farming and tenantry in this country,) he left. In a short time thereafter several columns of stuff appeared in the Chicago Tribune, purporting to be the result of an interview with me. You would have learned further that very little of the imputed interview was true as applied to or quoted from me, and also learned that nearly all exculpatory of you and in explanation of your manner of dealing with tenants, as I had learned in Lincoln, and explanatory of your methods had been omitted from the interview.[30]

Oglesby was responding to a complaining letter that Scully had written to Koehnle and Trapp. Oglesby was harsh in his criticism of Scully, who had not taken the trouble to investigate the truth of the *Tribune* story before writing critically of the governor. Oglesby's long letter suggests that Scully was ready to believe the worst and was accepting as truth the published accounts concerning the governor, without verifying them, just as many people were responding to stories that were told about Scully. Oglesby, on his final page, reminded Scully that he was different—alien—when he said:

> The liberal policy of this country in inviting to our shores citizens of all other countries was well understood both at home and abroad to mean encouragement to all those who sought homes and assylum [*sic*] here to come and occupy and cultivate the soil and enjoy freedom with us, such as it was and such as it might become. It never was intended or understood in any other sense, although a liberal law doubtless was in many instances perverted to shield a few who sought our country, not from love of its institutions, or respect for its people or character, but solely for personal gain, profit and benefit; whose "sagacity" enabled them years ago to see clearly enough through the atmosphere of cupidity the certain gains sure to come to all who would mass its cheap and rich acres into individual ownership. Others might improve the country and its morals; those important elements never disturbed the rapacity of those who took no care either of the welfare or the future of the Republic.

Apparently, Scully and Governor Oglesby were well enough acquainted to permit such a frank exchange, for Oglesby addressed the letter to the landlord's London home. In another context there were suggestions about Scully's use of politics in Illinois. That was in the essay "Alien Landlordism in America," by John Davis, one of the intellectual leaders of Kansas Populism who served four years in Congress and was editor of a Junction City newspaper. Davis reported:

> A single remark made by Mr. Scully respecting the Illinois legislature throws a flood of light on these points. He never fails, it seems, to be present during the sessions either in person or by attorney, and boasts that the legislature is always friendly, and has granted every favor that he has asked, and that he can evict a tenant more quickly and cheaply in Illinois than in Ireland.[31]

Davis acknowledged that "it is true, the Illinois legislature has passed laws forbidding the further acquisition of lands by aliens, but they did not affect former holdings." He explained further that "Mr. Scully favored these laws, because they tended to silence the public clamor against him, and caused him no inconvenience."

While Scully's nonresident and alien status was a big legal issue of the 1880s and was met with restrictive legislation, "no feature of the landlords' policies in the seventies and eighties was more disliked than cash rent."[32] Cash rent greatly reduced administrative expenses and required less supervision, so some landlords made use of this rental practice. However, share rent was much more typical on midwestern farms, with the usual division of one-third to the landlord; one-third to the provider of seed, fertilizer, implements, and draft power; and one-third to the tenant for his labor in working the land. When conditions were relatively stable or improving, cash rent drew few protests and was probably less expensive to a tenant than share rent. But if prices were declining or if the crop was severely damaged by climatic conditions or lost because of illness of a tenant, cash rent was exceedingly difficult to pay. Tenant unrest because of cash-rental practices was more evident in hard times—the mid and late seventies and the late eighties through the mid nineties. Associated with the criticism of cash rent was the Scully policy of taking liens on a tenant's crops, machinery, implements, and teams.[33] Sometimes the harshness of the cash rent was softened when the landlord abated part or all of the rent. Conversely, he could not increase rentals within the life of a lease contract, when tenants had unusually good crop conditions or received relatively high prices for products.

The Chicago *Morning News,* in a long analysis of Scully's operations in 1887, said that compared to the value of his farms, the rents were "a very moderate price," with typical amounts of $2 to $3.125 per acre plus taxes of 40 to 50 cents per acre. Neighboring farm land rented at $3.50 to $5.00 per acre; so the *News* believed that

> the hostility felt toward Mr. Scully is not due to his being a rack-renting, extortionate landlord, such as the Land League denounces, for he is not that, although some jump at the conclusion that he must be and rail at him accordingly. But he is very unpopular because of the system which he is introducing and maintaining. That system, however leniently it may be put into operation, cannot fail to blight any farming section where it is practiced for any length of time.[34]

Other contemporary analyses of cash rents on Scully's and neighboring land shows that Scully's rents were usually a little lower and were comparable when taxes and improvement costs were included.[35]

Henry George, an ardent backer of limiting land ownership through a single tax that would confiscate large holdings, held that the American prejudice toward landlords was "vulgar and irrational."[36] Perhaps that was the basis for the opposition to Scully's requirement that the tenant pay the taxes and assessments on the land. Supposedly, local improvements in every area of large Scully land holdings for such public work as schools and roads were "deplorably bad because the tenants do not want to pay taxes for repairs."[37] It would have been impossible to convince anti-Scully agitators of the 1880s that Scully's rents plus taxes were no higher than rents on other farm lands. They preferred to believe that tenants were burdened with taxes and assessments on the land and that Scully was thus avoiding taxation.[38] Somehow they had a vision of Scully basking in the sun in southern France or living high in London because his tenants were paying taxes in Illinois, Kansas, and Nebraska. Later Scully would say that

> the adoption of the measure [in Illinois] which compelled me to pay my own taxes was a fruitless measure. My tenants were paying the taxes. I made their rents so low they could afford to pay them. When the state compelled me to pay them myself I simply increased their rentals that much, which made no difference to me or to my tenants except that it removed the tenants beyond paying the taxes a practice which was [useful to those] who wanted to become familiar with the subject of taxes.[39]

Although Scully's American lands were rented for terms of up to five years, the landlord favored a one-year lease as soon as possible. Even though the one-year limitation was renewable to most tenants almost automatically, it drew harsh criticism, especially when coupled with the landlord's failure to provide a house and other farm buildings for a tenant. The conditions brought on by such a landlord omission were

> poor homes, poor schools, a sense of uneasiness, insecurity of property, and lack of independence. . . . There is little doubt that the Scully agents can control the 250 votes of their tenants in Logan county whenever they please to exert their power over them to do so. Inquiry shows that they have never done much in that direction, further than to occasionally request tenants to support some local candidate who is a particular friend. The system is, nevertheless, considered greatly pernicious by reason of the power it permits in that direction.[40]

The most evident feature of the Scully lands in the 1880s and 1890s to some observers was the poor quality of buildings and other improvements. In Illinois the Scully region in Logan County was described as a blighted area amidst "a thriving, prosperous farming community." Houses of Scully tenants were mere shanties, built for $150 to $200, with the whole set of improvements worth from $300 to $500. Tenants owned these improvements, and according to the lease, they could sell them when they moved. This pattern, accompanied by the insecurity of a one-year lease, was responsible for lack of "a reasonable development of the county."[41]

Scully's purpose in instituting a one-year lease was to improve the quality of his land by rotation and the planting of certain crops and to gain a better tenant. He did not want to keep a delinquent tenant any longer than necessary, whereas he willingly renewed leases when the rent was paid promptly and when a tenant tried to carry out the provisions of the lease. Most of his land was unimproved when he got it, so Scully saw no reason for providing buildings for each of his farms. Instead, it seemed better to him that the tenant provide his own improvements. Years later, Scully answered a question about why he did "not make improvements upon farm lands" by saying:

> I could not attempt it. With so many farms I would be at the mercy of an army of mechanics. I prefer to rent the lands at a price which will enable the tenant to make his own improvements. I always build foundations for buildings for him. They are a permanent improvement. I brick up cellars and wells and cisterns. I have put $350,000 in tile on my lands in Logan County alone. But I have not built houses and barns. I have put it within the means of the tenants to build these things himself. He can always sell them to his successor if he leaves the farm. If the house is his own it will have the effect of making him take good care of it. It will fit him for owning his own home on his own farm when he shall become a land owner.[42]

With only a few exceptions, Scully had an inviolable rule against selling land. Land was actively sold in the period 1855 to 1857, Ballycohey was sold in 1868 at the entreaty of Charles Moore, and Governor Oglesby bought a quarter-section in Sangamon County in 1888. Small tracts were sold to railroad companies, town-site companies, cemetery associations, church organizations, and school districts, always with the specification that the land would revert to Scully or his heirs when the purposes of the sale or when the organization purchasing the land were no longer served. Many efforts to buy land came from tenants who had saved the necessary capital or from other interested buyers who believed that Scully had so much land he would not miss a few acres; these efforts invariably brought the reply that an unwritten rule for Scully was "Sell no land!" Money was no object; the landlord was merely unwilling to reduce his holdings. The usual sale of land every generation or so did not seem to apply to Scully land areas, and it was criticized by many.

Near the end of the nineteenth century a typical Scully lease was lengthy, due to stipulations that many duties be performed or that the tenant be subject to a fine. Such coercion bred ill feeling even when the requirements were practices that a good farmer would follow anyway. These "burdensome requirements" were said by some to be exacting, and they could not be broken in court: one author described them as "iron clad, double riveted, with holes punched for more."[43] In general, these stipulations were hardly noticed by most tenants. "The exacting provisions of the Scully lease [were] not carried out to the letter as a rule";[44] nevertheless, the lease provided another area of complaint about Scully and his operations.

William Scully was shy and reserved and unwilling to face potential hostility; therefore, he dealt with his American tenants through his agents. As a member of the Irish landed class, he had an Old World feeling about class and its responsibility. He had had enough trouble in Ireland when he took over the role of the landlord's bailiff, and in the United States he sought to protect his position by a policy of avoiding tenants and not making their acquaintance. Similarly, he had very little contact with most of the people in the different communities where he owned land. It was his style not to make any effort to cultivate community good will and not to seek publicity for his operations. But even when people tried to find something of interest to tell about this "largest farm owner of the day," they were rebuked or obstructed at almost every turn. The power of his agents led to complaints that they were arrogant and that a tenant would not get his lease renewed if he crossed an agent. One agent was spoken of as "His Grace of Scully" because of his lordly mannerisms, and another was remembered for ordering tenants to doff their caps in Mr. Scully's office.[45] The landlord's distance from his tenants and his lack of communication with them created some of these stories, and his heavy dependence on agents and the power that he bestowed on them were responsible for others. To onlookers the Scully system possessed attributes that were decidedly un-American.

Other criticisms were directed at William Scully because of the quality and character of the tenants on his land. When the complaints of Scullyism were at a peak, his tenants were described as Germans, Danes, and a few Irish or native-born Americans, or "as ignorant foreigners, Bohemians, Scandinavians, and Poles" who made undesirable neighbors.[46] One newspaper proclaimed that "The Lord Scully tribe of aliens will have to go—so far as Illinois is concerned," and another spoke of Scully's tenants as a "'dreary and woebegone' lot of 'scarecrow tenants' who 'are in a state of absolute serfdom under his heartless alien rule, mostly transients' raising nothing but corn, year after year, from the same ground."[47] Kansas and Nebraska newspapers voiced similar sentiments about Scully tenants. Scully was said to have preferred hard-working farmers who were accustomed to severe self-discipline and were willing to accept a low standard of living. The style of lease and the pattern of overseeing the Scully estate attracted the poorest farmers around, according to some critics, whereas William Scully later declared quite the opposite. He claimed that the "Scully plan gave" his tenants "an opportunity to lease a farm when they had no money. What they did have, though, was more than money—it was a reputation for industry, earnestness, temperance and honesty with the Scully people.

So the Scully lease became a stepping stone to higher things, rather than a hurt to society. A young man finds it the means by which he can possess a farm of his own."[48]

A study of conditions in Hanover Township, Gage County, Nebraska, for 1885, shows a high standing for the seventeen Scully tenants there as compared to other farmers in the township. Scully's tenants were treated "somewhat liberally" in contrast to other tenants in the township. In fact, the Scully "tenants appear as a distinct group of farm operators who enjoyed the highest economic position in the township. The average farm acreage of this group was more than twice as much as that of others including owner-operators. . . . [while] the proportion of improved land was lower than that of other farmers. But, the value of farm capital and production of the farms on the Scully land exceeded that of other farm operators."[49] The "image of poor tenants exploited by a foreign landlord" did not hold "in the case of Hanover township."

Perhaps it was to his credit that William Scully did not employ a plan of public relations to acquaint a suspicious public with his goals of conservation for his land. On occasion he contributed to worthy community enterprises and to charitable activities, but these actions were never publicized. To newspapermen, the Scully in their stories was a "veritable demon, who grinds and tortures his tenants until they are nothing but miserable, groveling, poverty-stricken slaves."[50] When not given to such extreme language, newspaper comments contained a good deal of truth about the Scully land system. But similar complaints could have been directed to the problem of tenancy elsewhere on both large and small holdings. Scully gained great notoriety because of his visibility and because of his alien absentee cash-rent procedures, which required the tenant to provide improvements on a one-year lease. Contemporaries in the clamor against Scully in the 1880s, such as the Logan County superintendent of schools, saw no "good reason for charging Mr. Scully himself with being tyrannical and oppressive, or that his agents are, either." Most of the objections against Scully were "not as forceable as many claim," and "tenants do not complain as a rule." A neighboring landowner thought the "objections should lie against Mr. Scully's system rather than against his manner of operating under it." He also reported that Scully "rents his land at lower prices than he could get" and that he "is not grinding on his tenants." However, the Lincoln, Illinois, mayor "denounced the Scully system in vigorous terms." He believed the Scully tenants grew to be irresponsible and that their children were apt to become shiftless. To him, "poor roads, poor schools, poor citizens, poor homes, and shirking of fair taxes and honest debts were the legitimate results of the Scully system."[51]

A cynical appraisal by the agrarian radicals and reformers who sponsored the nonresident antialien restraints on acquiring land would hold that in "seizing upon a popular issue like Scullyism and riding it for all it was worth," they were "neglecting more fundamental issues."[52] Perhaps these reformers were demagogues; there was too much at stake to get a thorough revision of the entire system of public land. They used their energy, instead, to get legislative restraints on landowners such as Scully, which was certainly a side issue of that time. All that these new laws were likely to do was to stop Scully from making additional land purchases. If he obtained American citizenship, as Governor Oglesby was suggesting he should do, these antialien laws could not apply. As the 1890s opened with growing Populist sentiment in the three states where Scully owned land, he considered carefully the next steps for maintenance of his vast estate, which he wanted to pass on to his heirs.

wswswswswswswswswswswswswswswswswswswswswswswswswswswswswswswswswswswswswswswswswswswswswswswswswswswswswswswswswsws

9

The Fruition of Dreams

wswswswswswswswswswswswswswswswswswswswswswswswswswswswswswswswswswswswswswswswswswswswswswswswswswswswswswswswsws

By November 23, 1891, when he reached his seventieth birthday, William Scully could see that the vision of his great landed estates was reaching fulfillment. He had three young sons, who were absorbing his philosophy of long-range goals for the land, and a wife, then in her late forties, who was fully committed to his views. He had other plans for his youngest daughter. With a little more time, he could bring to culmination the disciplined goal of his adult lifework. He was given almost fifteen more years than the traditional three score and ten. He used these to provide the capstone for his dynastic empire of land.

But even a strong-willed multi-millionaire would not have everything his own way in his twilight years. As far as his estates were concerned, yes, he could lay out the policy under which they would be operated, with expectation of a high degree of compliance. But he could and did have strong-minded children, whose contrariness provided him discomfort and embarrassment. Even harder to fathom for landlord Scully were the views of "fickle" Americans who were "envious" of his foresight and good fortune. Earlier they had encouraged him to make his initial investments in American land. Now they were responsible for throwing roadblocks in his path to ultimate success, and he and his agents worked to minimize these obstacles. Personal tragedy also haunted William Scully in what should have been his happy golden years.

The 1890s in American history have been called the "Gay Nineties" by sentimentalists. Others have seen the period in a color sense as the "Golden Nineties," "The Mauve Decade," or the final years of the "Brown Decades," a part of the American Victorian era. One historian called it an "Age of Excess," and another defined the period as "The Restless Decade," separating the Old America, which was gradually disappearing with the passing of the frontier and with the new rise to world power for the United States. Strong indications of that decade brought an assurance that industry would soon triumph over agriculture in economic and political power.

There is no sign that William Scully saw or was willing to recognize the existence of these momentous changes in American life. However, his brief recorded memoranda of that time show that he was bitterly resisting the transformation of British institutions. He endeavored to cling as long as possible to his Irish estates under the older patterns of landlord dominance. His ties to traditional law and legal authority caused him to reject the new in favor of the old and trusted ways. Yet, there were many changes that William Scully endorsed. For instance, the early trips to Illinois had been by stage or horseback or lake steamer—railroads were a big improvement. Transportation of virtually every kind had brought greater comfort and speed since his first trip to America forty-one years earlier. Techniques for a more scientific agriculture had shown much improvement, and farmers better understood the new ways and were beginning to reap the benefits. Institutional arrangements for transferring large sums of money across international borders had much improved, and Scully was making use of them.

By the 1890s, with his western lands coming under lease, William Scully's tenants numbered more than a thousand. Requirements set down in the lease, which have been "laughed to scorn by frontier farmers," were now easier to enforce, since the tenants' alternatives in the way of other land that they might rent were rapidly disappearing.[1] In a Kansas lease of 1893, the tenant agreed to

> cultivate and manage said land in a good and husbandlike manner, that he will pull out clean out and destroy all burrs, thistles and other weeds on said land by the first of September in each year. That he will take care of, cultivate, protect and maintain all hedgerows, fences, fruit and other trees that now are, or may hereafter be planted on said land. That he will trim all hedges on said land by the first of January in each and every year during this lease and burn the brush. That he will at his or their own expense, keep open, cleanse, plow, scrape and dig out all ditches and drains that now are, or may hereafter be made on said land, by the first day of October in each and every year during this lease; and in case of failure to keep open, cleanse, plow, scrape and dig out said ditches, trim said hedge-rows, and pull out and destroy the burrs, thistles and other weeds, respectively, as aforesaid, the said Tenant agrees to pay said Landlord Seventy-five cents per rod for the

> ditches, twenty-five cents per rod for the hedges, and two dollars per acre for land in burrs and weeds, as damages for such failure in addition to the rent hereby reserved, such damages to be recoverable by the said Landlord in the same manner as rent in arrears. That said Tenant will not permit or suffer cattle or other animals, to feed upon the stalks standing on said land, said stalks being reserved to the Landlord; and that he will deliver up said premises to the said Landlord in good order and condition as they now are, at the end or other sooner determination of the period for which the same are let, reasonable wear and tear only excepted.[2]

Such a requirement was typical of all Scully leases of that time. Cash rent was also usual, but sometimes share rent was collected. For example, a Logan County tract of 144 acres had been unrented in 1889, so it was leased for "grain rent"—one-third of the small grain and two-fifths of the corn—in 1890 to attract a tenant. Whereas cash rent in 1888 had been $470.00, plus $56.86 paid in taxes, the share rent in 1890 was merely $404.26. But the next two years it rose to $810.92 and $853.10. Much more supervision was needed in handling share-rent contracts, and Scully preferred the cash-rent pattern. By 1900 most land in Logan County was rented at rates ranging from $5.50 to $6.25 per acre plus taxes. At the higher rate the 144-acre farm in 1900 would have produced $890 plus taxes.[3]

Income from Kansas and Nebraska land was far less per acre than from Scully lands in Illinois at that time. Scully's explanation for the difference, which did not take into account the nature of the crops and the more intensive Illinois agriculture, was that "settlement of Illinois is older, the farming more scientific and the railroad facilities better."[4] The high rent on Marion County land in the 1890s was $1.50 to $2.00 per acre plus taxes, whereas rent on Nuckolls County land did not average a dollar an acre until 1898 and 1899 when 6 1/2 percent of the total was returned to tenants in abatements and almost 13 percent more was lost by the landlord because of the tenant's inability to pay. Rentals in Butler County were about the same as in Marion County, while land in Marshall and Gage counties brought slightly higher rates.

The dry years of the 1890s, coupled with a farm depression and the Panic of 1893, brought dire prospects of rentals, especially in the western lands. The agent in Beatrice wrote Koehnle and Trapp early in 1893 that one tenant had

> skipped out for Oklahoma Saturday night and left his improvements for me to get his rent out of—I will probably get 400 out of $665—If I can get an honest man on the place instead of a [rascal] I will feel that the loss will not be so much. If the change in adminis-

> tration has any thing to do with the present depression I am not in favor of any more changes. However I think a good crop year makes the collection of rents easier than any other condition.[5]

Collections that year were down. In September the agent expected a 95 percent collection rate in Marshall County, while tenants in Nuckolls County returned 75 percent of their rent, and those in Gage County returned only 65 percent. Two months later the Beatrice agent reported:

> Collections are going to be much worse than I anticipated in Gage co. Nearly every man that comes in wants me to carry him over. Where they were expecting 15 and 20 bu. corn they are getting 8 & 10 and at the low price of 23 cents. What little they have to sell if any does not amt. to anything. A good many will have to buy corn to feed their stock through.
>
> What in *Hell* am I to do?
>
> Some of the tenants are proposing to flee the country in the spring and all I can possibly get will be the improvements on their leases. If it were an ordinary year I could sell the improvements for as much as the years rent but you cant sell anything here now for money. I have never seen anything like such a time since I have been doing business for Mr. Scully.
>
> Well, I will do the best I can and take anything and every thing I can get, except the women and children, and household goods. I may get better results than I anticipate, but it looks blue now.[6]

The drouth of the mid nineties in Nuckolls County resulted in seventy-five Scully tenants running out of "seed or grain to feed their teams to raise a crop with." Efforts to alleviate poverty conditions in Nuckolls County resulted in some landlords providing fifty bushels of corn to their tenants for each team they farmed with. The agent calculated that Scully's outlay, if he agreed to such a proposal, would be two thousand bushels at a cost of fifty cents per bushel. The Nebraska agent closed his letter to the chief agent in Lincoln with this plea: "Now the question for you to decide is whether to chance this much money, or let more or less of the land lie Idle. As time is pressing please telegraph your decission [*sic*]."[7]

By the 1890s Scully agents were usually expected to work full time for the landlord and not to engage in any other business. Lincoln, Illinois, continued to be the headquarters of the American estates, and Frederick C. W. Koehnle and Frederick Trapp were the landlord's chief employees. Other agents had limited authority and dealt with the landlord through the Lincoln office. For example, the only written instructions for F. William Fox, Scully's agent in Marion were:

> I authorize F. W. Fox, of Marion, Marion county, Kansas to receive all my rents, and to make and enforce all collections for me in Marion and Dickinson counties, and to sue for the same whenever he himself shall think it advisable to do so; and to take other proceedings for the recovery of same as he may be advised to do so by Fr. C. W. Koehnle, of Lincoln, Illinois, or other of my duly-appointed agents in the United States.[8]

By the 1890s a set of "Rules on Scully Farms" had been worked out as a guide for the kinds of tenants sought by William Scully. In a paternal fashion, copies of this list were distributed to tenants. They said:

> Tenants must be temperate.
> They must pay their debts.
> They must not quarrel with their neighbors.
> They must build their own houses and barns and plant their own orchards.
> They must make money for themselves over and above the money with which they pay their rents.
> Their stock must be well fed.
> Their machinery must show good care.
> They must so deport themselves that the community will respect the Scully colony and the name of Scully.[9]

In the meantime, in the depression year of 1894, Scully's agents began to purchase land in Bates County, Missouri, which is located on the Kansas border, two counties south of Kansas City. In the face of a hostile league of landowners who agreed to withhold sales to Scully, initial purchases that first year amounted to thirty thousand acres, which were paid for with cash. Rumors persisted that Scully was making his purchases in specific sections radiating out from Butler, the county seat, and that he was not buying land in between. A conspiracy theory was advanced that the newly acquired "land is to be stripped of buildings and colonized with Italian labor of the cheaper and more degraded sort," so as to make the locality undesireable for adjoining landowners who would then sell cheaply to Scully.[10]

Purchases continued through 1895 and 1896 at an average price of $27 to $35 per acre until Scully had 41,844 acres in Bates County. Some of the land sellers became tenants on land that they had formerly owned. Other new tenants were primarily Bates County farmers, thus laying to rest the hearsay reports of strange, foreign farmers who would be imported to lease the land. Butler became the site of a new Scully agency, and rents were set at a per acre rate from $1.50 to $2.50. Scully's new estate gave him about one-thirteenth of the farmland in the county, for which he paid about one-tenth of the taxes, a fact that led the Scully agent to inquire of the landlord what the tenants should do on the vote for a new county courthouse. Scully responded by saying, "Vote for it," and the issue passed.[11]

The outlay for the Missouri part of the Scully estates was roughly one and one-quarter million dollars, which was equivalent to the total expended for earlier Scully acquisitions in America.[12] By the end of 1896 William Scully was the owner of American territory equal to nine and three-quarters townships, or more than 351 square miles. His total was 224,738 acres of farm and pasture land in the four midwestern states of Illinois, Kansas, Missouri, and Nebraska. Only minor changes in total acreage came in subsequent years. His expenditure for this land, purchased over a period of forty-seven years, was somewhat under three million dollars, perhaps only a little higher than two and one-half million dollars. Scully land in Illinois was found in seven counties, with most of it located in Logan, Grundy, Sangamon, and Livingston and very small acreages in Mason, Tazewell, and Will.[13] The Kansas counties were Butler, Dickinson, Marion, and Marshall, with the major concentration in Marion. Two-thirds of Scully's

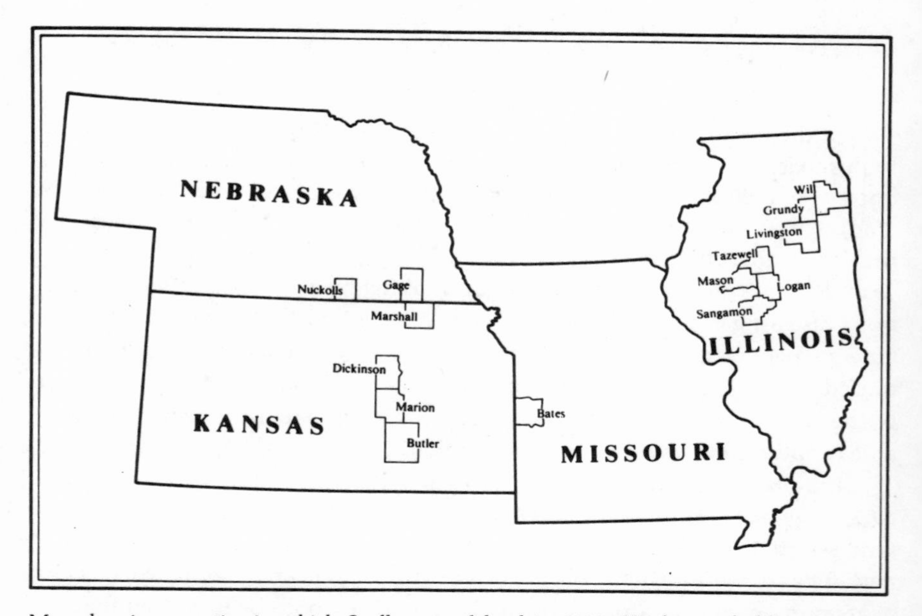

Map showing counties in which Scully owned land in 1900. His largest holdings were in Marion County, Kansas—55,666 acres; Bates County, Missouri—41,884; Nuckolls County, Nebraska—40,830; Logan County, Illinois—34,728; and Gage County, Nebraska—23,157. Courtesy of *Western Historical Quarterly*.

Nebraska land was in Nuckolls County, with the remainder in Gage. Bates County, Missouri, became the fourteenth county in Scully's landed estate in America.

On September 20, 1895, William Scully appeared before the clerk of the United States District Court, Southern District of New York, to initiate his naturalization as an American citizen.[14] Newspaper commentary in later years generally interpreted this move either as a response to Populist party hostility and the legislation against aliens that they were able to pass or as the landlord's reaction to events or happenings that came later than the date of filing his declaration of intention. Most of the Populist activity came after the passage of the antialien laws. Certainly, the legislation of the late 1880s and the early 1890s had some influence on Scully, but two developments, very close in time, probably confirmed his resolution to take that step toward American citizenship. The first was the death of his son William, at the age of eighteen, probably in June, 1895, in the South of France. The death of this son, the eldest of Angela's children, hit the landlord hard, as he had great expectations for the young man. Because "Willie" wouldn't be around to take over and because William was less sure that Thomas and Frederick would have mature judgment as landlords, he may have felt pushed to renounce his British citizenship in order to protect his property. A mention five years later that grief for Willie still prevailed in the home of William Scully tells only a little about this loss.[15] Even though a Supreme Court decision supported alien ownership of land, a second development that may have influenced William Scully to seek American citizenship was the passage of a Missouri law, in 1895. This hasty action came at the very time when he was acquiring land in Bates County, and it prohibited the acquisition of land in Missouri by nonresident aliens.

Scully wrote to his agents in Lincoln that he had made a declaration of intention to become an American citizen in 1853, and they found it on file in the Logan County courthouse. Since more than forty years had elapsed, Scully received legal advice that it was better to start over to provide continuity for the five-year waiting period required for naturalization. This move and the subsequent establishment of the William Scully residence in the United States ended for all time the possibility that William Scully would be elevated to the British peerage, which was supposedly one of the fondest of Mrs. Scully's hopes.[16] It also served another purpose in contributing to greater cordiality in American newspapers and in greatly reducing criticism of Scully landlordism. No other action taken by Scully in the nineteenth century found such widespread public support. That he was serious about gaining his American citizenship can be seen from two small bits of evidence. Scully was on the voter's list in his London district in 1890, but not thereafter.[17] Before his five-year waiting period for citizenship had been completed, he asked those few agents who were not American citizens to take out their naturalization papers.[18]

News reports about William Scully's intention to become an American citizen were speculative and inexact on a number of points. For example, a statement was made that Scully had sold all his English and Irish property that was not encumbered. Also, the location of the American residence for the landlord was as yet undetermined, but after staying a while in Lincoln, Illinois, it was expected that he would settle either in New York or Boston.[19] In reality, Scully owned no land in England; his residence in London had always been in a rented house. The only land that Scully had ever sold in Ireland was Ballycohey. None of the Irish estates were sold at this time, and throughout the remainder of his life he resisted giving up any of his holdings in Ireland. Washington, D.C., seems to have been picked for a family residence because of the lack of citizen obligations in the District of Columbia and in order to avoid summons or taxes that might be imposed on a citizen of a state. It was there that Scully rented the old home of Gen. John A. Logan in Calumet Place on the outskirts of the city.[20]

A large staff of servants was employed to care for the Scully family in Washington. Carriages were purchased, footmen and coachmen were secured, and the family accepted some social invitations and "gave some entertainments." This entrance into society was "on a scale of elegance, but nothing to excite unusual attention." Most active socially were Mrs. Scully, who was described as "a handsome, matronly woman, [who] was well liked," and Ita, "a beautiful young girl of pleasant manners and inclined to athletics, [who] became popular in a quiet way." Reportedly, in 1901 she frequently assisted Mrs. Thomas F. Walsh with her entertaining. Neither Thomas nor Frederick cared much for society activities in Washington, and no mention was made of William Scully's participation other than a general statement, when he was eighty years of age, that he met all business and social obligations with the "energy and interest of a man in the prime of life."[21] While residing in the nation's capital, both Thomas and Frederick Scully took a course of study in the legal offices of William Scully's Washington attorneys, Jackson H. Ralston and Frederick L. Siddons. Their training was in "contracts, leases, deeds and the court processes involving these particular branches." But other activities were of interest to the young men in the family.

For instance, in 1900 Thomas secured a steam engineer's license to operate his steam automobile, a Locomobile. Thomas and Frederick knew that their mother greatly enjoyed picnics, so they asked her to go with them in the car to a picnic. She would have nothing to do with the car, but a picnic was fine, and she would go in her coach. So the boys headed out in the Locomobile to the picnic site, and after going a ways, the car stalled. Soon their mother passed them in her coach and haughtily refused to stop and pick them up. The boys finally got the car started again, and soon they passed their mother in a cloud of dust. She was furious at their discourtesy.[22]

About two years after William Scully received his citizenship papers on October 17, 1900, the Calumet Place house was given up. Another house on Scott Circle, at 1401 Sixteenth Street, N.W., known as the Gurley House, was purchased in 1902 at a cost of $90,000. Additional expenses on the house and the cost of furnishings brought the total to $101,596.24. This "poorly architected" house was rented to Gen. Russell A. Alger for a period of four years beginning January 1, 1903. Alger, a United States senator from Michigan, was a brevetted brigadier general from the Civil War, a former governor of Michigan, and secretary of war for two years under McKinley. In late 1902 the Scullys returned to their Holland Park residence in London, with plans to move back to Washington at the expiration of Alger's lease.[23]

William Scully's Irish estates in the late nineteenth century gave him net profits on his cattle and sheep alone of £1,769 per year from 1890 through 1900.[24] The Irish Land Acts of 1870 and 1881, sponsored by the Gladstone administration, greatly restricted the Irish landlord's freedom and provided a feeling that sizable agricultural properties would be seized by the government to satisfy demands of the land league. One report in 1887 speculated that Scully had kept his Irish estates "in trust or set apart for three daughters by his first wife."[25] But such was not to be.

In 1899, because of premonitions of governmental confiscation and in order to avoid death taxes, Scully sought to plan for the future of his Irish lands by giving his Tipperary estates to Thomas, his oldest living son, who had just reached his twenty-first birthday. This land had come to William Scully "for his lifetime only and was then to go on to his heirs male." He gave his wife all of the cattle, horses, and implements from the Tipperary lands, as well as all of his Kilkenny property.[26] Elaborate instructions to the new owners were written by William Scully because "the present land laws operating in Ireland . . . . are confiscating the landlord's interest." Scully pointed out that his "Kilkenny Estate . . . chiefly fit for grazing . . . [but not suited] for wintering stock" could be divided into six or eight parts and then be leased or sold. Instructions were given about the steps to be employed—first, to reduce the number of cattle; then to advertise the land; and finally, the procedures for taking payment. He cautioned that his Forgestown farm, which was about halfway between the Kilkenny lands and Ballinaclough, should be kept as a cattle resting place until the Kilkenny estate was disposed of. Then farming could be curtailed on his other lands, with Ballinaclough as his final holding ground.[27]

But Scully would never permit the sale of his Irish land. On one occasion, Thomas went to considerable effort to contact the Irish Land Commission and to reach a price for the Kilkenny estates. All his hard work brought an absolute refusal from his father—he would not sell. That land had come from his beloved brother.

William Scully must have felt that all of his children were irresponsible and immature. Thomas, because of his strong interests in disposing of the old family estate in Ireland and because of his temper, worried William most of all. Directions were given to his wife, as he turned over his vast estate to her, to use care in giving complete authority over the estate to her children. There were written instructions, which suggested that Thomas should receive a substantial reduction of a share in the estate, but final discretion was given to his wife.[28]

About the same time that Scully filed his declaration of intention to become an American citizen, general farming conditions began to improve. So the newspaper comments about the large-scale landlord began to moderate. The editor of the Marion (Kansas) *Record* reported that the landlord "erroneously known as 'Lord' Scully, has been in Marion the past week. Mr. Scully has never claimed any English title. He is a plain Irishman, now seventy-odd years of age, and, though he probably owns more American land than any other man, he is as unassuming in manner as the humblest citizen. He has recently become a citizen, so the laws against alien land ownership no longer affect him."[29] However, on the same day, the Lincoln (Nebraska) *Courier* was critical of Scully's lack of philanthropy in the Lincoln area, and a little later the Superior (Nebraska) *Journal* attacked the Scully administrative procedure in Marion County, but not in nearby Nuckolls County lands.[30] But generally, reports about Scully in his last decade suggest that he had mellowed, and more positive aspects of his huge estate were discussed. It was discovered that neither Scully nor his agents pushed his tenants for rent during poor crop years. From Marion County in 1901 came the report that Scully was "very lenient with his tenants."[31] Also, when Illinois farmers suffered an unusually bad year in 1903, Scully agents wrote the landlord and suggested that abatements be given for half of the rent. Scully said for them to forgive all of it.[32] Similarly, it was discovered that neither the "burdensome requirements" nor the one-year lease posed a handicap to serious farmer-tenants. Scully's insistence that "his agents keep abreast of the progress of agricultural practices and to urge the tenants to do the same" was heralded as progressive agriculture.[33] In anticipation of the alfalfa requirement on the Scully leases in Nuckolls County, the agent in Nelson ordered books on alfalfa, written by the secretary of the Kansas State Board of Agriculture, for each of his tenants. The landlord paid for the books.[34] Constantly, Scully urged his agents to deal honestly with tenants and to expect honest effort from them.

The new and more generous attitude toward Scully in the press prompted him to grant an interview to a young reporter of the St. Louis *Post-Dispatch* in 1901, which produced the only direct William Scully interview in an American newspaper. When the reporter visited Scully's office in Lincoln, he was told that "Mr. Scully was there, but was very busy and did not see newspaper men." The reporter asked the agent to announce him "and learn from Mr. Scully himself if it were impossible to see him." The landlord responded through his agent by asking for a list of written questions for which the news reporter wanted an answer. These questions opened the door, and Scully granted the interview for the next morning.[35]

The interview began with Scully saying: "Young man, I salute you as a friend. Your written questions have convinced me you are interested in me in no unfriendly way, and I shall be glad to be at your service. You have put some of these questions in a blunt fashion, but they are business-like and give no offense." Thus, the most complete newspaper story of Scully's life was obtained and published. According to a later story, one of the conditions under which Scully granted the interview was that he could see "the story before it was printed," an indication that he could not fully trust newsmen, however friendly they might seem.[36]

Throughout his career, William Scully handled all of his personal and business correspondence without a private secretary. The staff in each of the Scully agencies located near the main concentrations of land did grow. Generally there were two or more agents at each office, one of whom was older and more experienced in order to maintain continuity and knowledge of the Scully way of doing things. By 1904, on Scully's last trip to America, sixteen men posed with him in a picture. They probably constituted his entire agency staff in America. Four—Frederick C. W. Koehnle, Frederick Trapp, Jeff Sims, and Fred Koehnle—were from Logan County. Three were from Bates County: a Mr. Arnold, a Mr. Crowe, and Jesse Smith, who had been sent by Scully to the University of Missouri for specialized training. Captain Henry Fox was the Grundy County agent. All other locations were represented by two agents each—Henry Fox and W. W. Hawley in Nuckolls County; a Dr. L. P. Rogers and Ed Fisher in Gage County; John Powers and Billy Evans in Marion County; and John Cole and an unknown person in Butler County.

In 1905 Frederick C. W. Koehnle, as Scully's chief agent, was living in pretentious fashion in a big house in Lincoln. His salary was $6,000 per year, a figure that remained remarkably consistent during his long career. Trapp's salary, which had been $3,000 in 1899, was also $6,000. Other agents were in the $1,500 to $2,500 range. Even Thomas and Frederick Scully were on the payroll, although their duties consisted primarily of visiting the estates at various times to become acquainted with the land and the agents and to learn how to examine the books. Koehnle and Trapp visited the outlying agencies in the years when Scully did not come to America. During those same years, according to lists of travel expenses in their office, either Koehnle or Trapp, or both of them, would travel to England to give a personal report to the landlord. On one of his business trips to visit Scully, Koehnle was invited to bring his family, and they were treated to a Grand Tour of the Continent.

During the last seven years of his life, William Scully's account in the First National Bank of Chicago showed the following deposits:

Sept. 5, 1899–April 2, 1900	$275,000.00
August 9, 1900–February 25, 1901	262,853.32
March 1, 1901–May 13, 1902	195,500.00
September 4, 1902–March 28, 1903	263,097.46
1904	283,676.52
1905	267,000.00
1906	326,937.00[37]

These figures are probably an indication of his net income per year from his American estates, not of his Irish income or of his income from any bonds and other securities.

When John Scully died in 1885, leaving his widow and two very small children, he left a fairly sizable estate for a man approaching the age of thirty-six. William Scully helped Louise C. Scully by providing administrative services for the land that John had left. William enjoyed John's children, but he took no steps to be responsible for them. At the age of forty-seven, Louise C. Scully died on May 29, 1899, and William began to see that her children, who were teen-agers at the time, would need some guidance, even though Louise's family were helping out. To no avail, Frederick Trapp from 1886 on had urged William Scully to help John's children, reasoning that he could readily afford to do so. Finally, in Scully's old age, the Catholic priest in Lincoln talked to him about his responsibilities to John's children. Thus, William Scully, an Anglican for more than thirty years, responded by giving John Chase Scully and Louise Scully a considerable stake in the Scully estate.

John was given the poorest land in Kansas—the 9,058 acres in Butler County. Louise was given a much smaller amount of land in Illinois and, in addition, income from John's land to the extent of $3,000 per year.[38] John's gift was deeded on July 9, 1906, and was recorded in the Butler County Register of Deeds Office on January 7, 1907. Both John and Louise maintained close and cordial ties to William's family, and their association continued through their descendants.

William Scully's American will, two pages in length, was drawn up and witnessed in Washington, D.C., on April 2, 1901. Since his Irish lands had been transferred to his wife and to Thomas by that time, he stated: "I believe that I now own no property of any Kind, outside the U.S. America." He bequeathed his lands to E. Angela Scully, his wife. Before stating the nature of the guardianship for minor children, the executors, and their bond, he emphatically recorded:

> I leave my three daughters by my first wife, Mary and Julia, (both of whom, as I hear have been married,) and Kathleen, and to their child, children, and their descendents, and each of them, nothing whatever—My Will and meaning being absolutely and entirely to disinherit all, and each of the three of them, and their child, children, and their descendants—I do this for superabundant reasons, which I forebear to particularize here.
>
> I recommend—however—Kathleen, the younger of them, to the kind consideration of my dear wife, E. Angela Scully—But I leave my said wife absolutely free, and unfettered therein.[39]

The will contained no bequests for any of his children, even those by Angela, who were twenty-eight, twenty-six and twenty-four years of age when their father died. William Scully would let his wife take care of that.

In spite of the will, Scully deeded the land to his wife shortly before his death. News reports in August, 1906, stated that the registers and recorders of deeds in counties where Scully owned land had received one of "the most peculiar deeds ever placed on the books." William Scully "simply stated in substance that he deeded to his wife all the real estate in said county of which he might at the time be possessed." The comments on this action were that Scully was dropping some of his usual business care in this move.[40]

Resentment against Scullyism eased with growing farm prosperity in the early twentieth century. Some observers were identifying the Scully tenant system as a new rung in the agricultural ladder, enabling a tenant to start farming on a Scully lease with very little capital. The tenant, because of low cash rent, could put money into improvements that he could eventually sell in order to make a substantial down payment on a farm of his own.

After visiting his lands in 1899 and 1901, William Scully did not return until 1904. Then he spent the months of May, June, and July in Illinois and the other states on what would be his final visit to his American holdings. As on earlier occasions, he was accompanied by Thomas and Frederick. Independently, they visited the western states that year, earlier than did their father. Many changes had taken place since William Scully had visited these newer states more than thirty years earlier. He was proud of the vast estate that he had built and organized over a period of fifty-four years, and he tried to communicate that feeling to his sons, who would be taking over soon. He believed that the honest and forthright dealings with his tenants had paid off. If it was paternalism in his relation to his tenants and employees, he believed it was in a benevolent form. In his declining years his estates projected the fruition of a long-time dream. Had he given his sons a proper appreciation of the land so that it would remain in the family? That question must have risen in his mind many times as he built a series of safeguards into the bestowal of land to his wife. No doubt there were specific oral instructions, which he would permit her to modify as time would pass.

During his final years, when London was again his primary residence, Scully traveled with his wife to southern France to take the sun. Some time was spent in country houses in England, but most of each year found him at 12 Holland Park. Plans were under way to return early in 1907 to the Gurley House in Washington, D.C., when William Scully died at his London home on October 17, 1906, just over a month before his eighty-fifth birthday. The cause of death was cardiac failure, aggravated by fever and intestinal catarrh.[41] Although some news accounts said he would be brought back to Washington for burial, his body was placed in the Kensal Green Cemetery, located north of Holland Park. Later, a simple, large, uncut dull red stone was raised over his grave. On it was carved:

In Loving Memory
-of-
William Scully
Born 23rd November 1821.
Died 17th October 1906.
"Rest in the Lord"[42]

wswswswswswswswswswswswswswswswswswswswswswswswswswswswswswswswswswswswswswswswswswswswswswswswsws

10

The Scully Estates in Later Years

wswswswswswswswswswswswswswswswswswswswswswswswswswswswswswswswswswswswswswswswswswswswswswsws

Remembrances of William Scully emphasized his large tenant-operated landholdings, his use of cash rent for one-year leases, his unwillingness to supply improvements, and his various conservation measures. Unrealistically, these elements of the Scully system were treated as if they had always existed, rather than as the accumulation of experimental practices of a lifetime. Scully was also portrayed as a prophetic capitalist who combined realism and great vision; at the same time he was presented as the harshest of harsh landlords with absolutely no feeling for his tenants, one who imposed onerous obligations on those unfortunates who happened to farm his land.

As might be expected, some truth and some falsehood lie in all of these assertions. Even to assign to Scully great skill and shrewdness in carefully selecting his first holdings in Illinois, without recognizing that this was the only large block of land then available or that he soon decided to sell, ignores reality. Forgotten, in most stories of Scully's early years in America, was the fact that he had something like $85,000 to pour into low-cost land. This wealth would have evaporated or gone down the drain had he not had other income in Ireland to fall back on during the years that expenses on his American estates far outpaced his income from them. Also, this choice, high-quality land gained the superb classification after tiling and draining, which required heavy investments over many years. So, statements about William Scully—that he never sold land or that he always required one-year leases or that he could neither read nor write or that he never provided improvements or that he always found local opposition—become suspect, if not actually untrue. More reasonably, Scully may have recognized the inherent value of the wet lands that he acquired in Logan County in 1850 and 1851, but of more importance to him was getting that property at a very low price. Once he had the land, he did not know what course to pursue; first he started farming it himself; then he began to sell. When neither worked out to his satisfaction, he showed a considerable flexibility in finding ways of leasing his unimproved land. As might be expected, he relied heavily on his Irish background for usage related to leasing. Then he transferred almost all his personal attention to his Irish lands until his experiences at Gurtnagap and Ballycohey soured him on the future of Irish agriculture. Thus, when he was ready to concentrate his scrutiny on his American estates, time had been on his side, and the value of these estates had increased almost tenfold from twenty years earlier.

William Scully's long life was one of learning and growth and one that was not without problems produced by his own self-image. He was merely an observer from the gentry—the landed class—until 1843, when he was given some land in Tipperary. As a beginning landlord, from 1843 to 1850, he survived the potato famine and even saved money. During the 1850s and 1860s he got his start in Illinois and gained horrendous notoriety for his landlordism in Ireland. The prosperity of his American estates after 1870 and his expansion into Kansas and Nebraska gave him the opportunity to put into operation his ideas regarding the proper way of handling an extensive landed estate. Agents, whom he selected to get the job done, stayed with him for many years and usually they were good appointments. By 1890 the pattern, typically recognized as the Scully land system, was emerging. The days of trial and chance success were over. More wealth was available for adding lands in Missouri. Scully looked upon his younger family as the reason for an extension of his vast landed estate into the future. The years from 1890 to his death in 1906 were employed to assure the perpetuation of the Scully way of leasing land on his American holdings. In looking back over his long life as a landlord, William Scully had a tendency to see each step in the amalgamation and maintenance of his huge estate as a part of a premeditated plan, where in actuality a considerable number of them were accidental. He did not dwell on the problems and opposition that he had encountered. Instead he maintained:

> My landlordism in the United States has never injured society in any wise. I have been strict. I have been careful. I feel that I have been fair and that I have been honorable in all I have done. I do not believe you can find a man in any of the states in which I have land who will say that I have done an unfair or a dishonorable thing.

> Would it not be easy to know it if I had been an oppressive landlord? Would my tenants have stayed on my farms in this great land of free men and innumerable farms to be had for the taking?[1]

Of course, by the twentieth century his landholdings were completely under lease. He was then portrayed as the "nation's most extensive landholder."[2] His ideas about conservation of the land's fertility through the use of clover or alfalfa were being incorporated into each formal lease. His benevolent paternalism with his agents and his tenants was increasingly displayed. Contrary to speculation, he did not foresee a time when his lands would be divided into small tracts and sold outside the family.

William Scully's wealth at the time of his death late in 1906 was variously reported at figures from ten to fifty million dollars.[3] Land values in Nebraska and Kansas doubled between 1900 and 1910, while there were less spectacular increases in Missouri and Illinois. It would not be unreasonable to place the value of the estate that he put together in America at fifteen million dollars, plus additional amounts that could have been just as high for securities that he owned. These reports, of course, did not take into account the fact that Scully had transferred his lands to his wife and to others shortly before his death.

His will, which reinforced these transfers, bequeathed all of his American holdings to his wife. It was filed in the probate court in the District of Columbia on February 1, 1907, delayed to that time until an agreement, completed the previous day between Mrs. Julia Aungier, the middle daughter of William Scully and his first wife, and Mrs. E. Angela Scully, could be signed, in which Julia renounced all claim to her father's estate. While the administrative docket said that Mrs. Aungier would receive "$1.00 and other considerations," she was actually paid £5,000 sterling two days after signing and an additional £5,000 thirteen months after the date of admitting the will to probate.[4] Almost two years later, Mrs. Scully appeared in the court to answer questions about the estate, and it was settled almost a month later.[5]

E. Angela Scully retained ownership of the Washington house for all of her remaining years, and she lived there part of the time. But after 1912 she usually resided at 12 Holland Park in London. In keeping with the oral instructions from William Scully, she did not divide the Scully estates among her children for another dozen years. She did draft English and American wills in 1912, perhaps as a consequence of two serious bouts with gallstones, which required surgery that was successfully performed in 1915.[6] In her English will, Thomas was given additional Irish land, but he had already gotten his mother to sell most of the Kilkenny lands, for which a lower price was received than that negotiated when his father was alive.[7] In that will, most of E. Angela Scully's properties and personal effects in the United Kingdom were to go to her daughter, Angela Ita Harriet, who had become the wife of Edward Arthur Parry. The daughter was requested to "look after and provide for Kathleen Scully . . . as I have done in pursuance of a request" from William Scully.[8]

A considerable part of Mrs. Scully's wealth was invested in bonds and securities, using criteria set down much earlier by William Scully. The agents in Lincoln were responsible for buying, selling, clipping coupons, and handling these investments. The matter of potential English inheritance taxes on her estate began to plague her advisers. So Frederick Trapp rented an extra safety-deposit box for her in Washington in which she could put bonds and securities so they would be out of reach of the British government. Later, a safe, which weighed three tons, was purchased and housed in a Washington warehouse so that unregistered securities and other family valuables could be stored. Trapp explained that

> the object of renting this additional box is, of course, to prevent the English government from attempting to collect an inheritance tax upon your American estates, lands as well as bonds, by a seizure of the bonds which you intend to devise to your daughter, which would, as a matter of fact, take all of those bonds to realize the amount that would be assessed against the property in America.[9]

Under the ownership of E. Angela Scully there were no noticeable changes in the extensive Scully estates, even with regard to her sons, who continued to be carried on the payroll. Hers was a holding operation before passing the wealth to her children. Koehnle had died, and the Lincoln partnership brought in Pickrell to form the firm of Trapp and Pickrell. Annual deposits to Mrs. Scully's bank account in the United States in the remaining years of the golden age of American agriculture continued to be high. From 1907 to 1918 they ranged from $235,000.00 to a high of $404,317.24, for an annual average of $365,000.00.[10] When her annual income from all sources reached $800,000 in 1918, she decided to do something about it by dividing her estate. Thomas, her elder son, was given the land in Illinois and Missouri; Mrs. Parry, her daughter, was given a large block of securities; and Frederick, her younger son, was given the Kansas and Nebraska land. At the ages of forty, thirty-eight, and thirty-seven, her children, she thought, had gained the proper maturity to be entrusted with great wealth, and they could be trusted to control it in the Scully way. On the advice of Frederick Trapp, she divided her gift to her three children, made at Christmastide, 1918, into equal thirds. Later she would follow this practice of not favoring one or two of her children over the others.[11]

By the time John C. Scully received the Butler County land just before William Scully's death, he was living in Peoria, Illinois. His schooling had been climaxed by a law degree from the University of Michigan, and he had established a law practice in Peoria. His Butler County property was handled separately from the lands of E. Angela Scully, but his agent still communicated with agents for the other Scully lands. On their biennial visits, Thomas and Frederick Scully were sometimes accompanied to Kansas by John. All the Scullys were captivated by the oil discoveries in Butler County, beginning with the spectacular strike at Stapleton No. 1 in October, 1915. Some of John's land quickly came under oil leases, and successful wells were drilled on it. So, oil royalties added to the bounty that he had received from William Scully.[12] Louise, his sister, had married Clark D. Simonds and had moved to Portland, Oregon; but her income from land in Butler County, Kansas, and from some land in Logan County, Illinois, continued.

Between 1914 and 1917 Frederick Scully bought about 27,000 acres in Lafourche Parish, near Cut Off, Louisiana. This land, which was located about 35 miles south-southwest of New Orleans, was mostly swamp and needed drainage before it could be used for farming. Eventually, about one-tenth of this area was set up as a private agricultural experiment station and was named Clovelly Farms Plantation. Many years were to pass before these Louisiana investments were to produce revenue. Frederick's assets in Louisiana, however, were not limited to products of the land surface, for oil was discovered on this land before it was found on any of his other properties. Try as he might, he could not strike oil on his Marion County land, even though there were producing wells in adjoining sections. Thomas was so proud of the tile drainage system in Logan County, Illinois, that Frederick began such a project in Gage County, but it was soon abandoned as being entirely impractical.

Thomas Scully continued to oversee the farming on his Irish lands. Like his father before him, Thomas went armed when walking about Ireland, but not with a revolver hidden in his pocket—he openly carried a Winchester carbine. Later he commented that they had expected to lose their Irish estates. "We all knew that we were living under the sword of Damocles for in those days as now there was considerable insurrection in Ireland." Because of the pressure for sale of large private holdings, except for only a small acreage containing an ancestral homestead, such as Ballinaclough, "all of the Irish lands had to be sold after the passage of the Irish land reforms." The price that Thomas Scully received was far below the value that he placed on the land, so he went to Ireland to appear before the Land Commission. He knew that there was little chance to get the price he wanted, but somewhat proudly, he later recalled that "I accused them in the very stronghold of their knavery and received a judgment of 2,000 pounds more."[13] However, the Irish Land Commission's final offer in 1925 was less than the price that the land would have brought in William Scully's time and was only about one-fourth of what Thomas thought it was worth. Moreover, the payment was in Land Bonds salable at 90 percent of face value, which caused Thomas to sell them the day that they were issued.[14]

During the years that E. Angela Scully owned the vast Scully estate there is no record of her having visited each of the counties where she had land. However, the requirement for maintaining alfalfa on a specified portion of each farm was extended to all leases in Nebraska and Kansas. Not until 1918 were all Scully farming leases in those states growing alfalfa, a crop that was considered so sure that the rent would always be paid. Clover was to serve much the same purpose on other Scully lands, and the proportion of each farm that was sown to clover was gradually increased. Both alfalfa and clover, as legumes, would add to soil fertility; but at first there was tenant resistance to growing them. However, the recognized worth of these crops, plus a proviso in the lease that rent would be abated on the portion of the land on which legumes were grown, produced strong tenant support. The Illinois leases also stipulated that ten acres in bluegrass be maintained as permanent pasture on each farm.[15] Mrs. Scully's contact with her estate was primarily through Frederick Trapp. She carefully watched expenditures and helped to establish policy on the estates in the manner that she had observed when the lands were owned by her husband. Estate matters took little of her time, and most decisions were made without her direct knowledge.

In the final years that William Scully owned the land, he asked his agents to report the number of planted trees on each farm, and he supported efforts to increase their numbers. In Nuckolls County, Nebraska, few of the Scully leases had many trees, and sixteen of them were reported to be "entirely destitute of trees" near houses. Thus, tree contracts were prepared to compensate tenants for planting and maintaining trees, somewhat as had been done earlier in Illinois. Mrs. Scully continued to get these tree reports during her ownership, but agents claimed that few tree contracts were in force.

Both Thomas and Frederick Scully promoted additional use of legumes on their lands immediately after they gained ownership. Tenants generally prospered in the years during World War I with the increased production and higher prices for farm products. Cash rents had not kept pace with the increased revenue that was generated on each farm or with the value of the land. So, late in 1918, the one-year leases prepared for the following year showed a substantial increase in cash rent. About that time the war ended, and prices for farm products stayed up briefly, and then collapsed. The

Scully tenants began to chafe under the pressure of the new higher rents in the face of declining farm prices, and a revolt was brewing that was quite different from the troublous times of the 1880s, when outsiders had been the cause of Scully's grief.

Dissatisfaction broke out among Scully tenants in Illinois in 1918, just before the transfer of the Illinois estate to Thomas Scully. A later interpretation of this friction suggests that it involved a struggle for power between Henry Fox and Frederick Trapp, now partners and chief agents in Lincoln. Fox was trying to precipitate an uproar in order to get the Scullys to oust Trapp, then in his late sixties, who had been with the estate since 1886. Trapp warned Thomas Scully about taking action to stir up tenants, because, he said, "Your father and I went through a tight spot one time and agitators went to the legislature and got a law restricting descent of land for alien absentee owners which forced him to become a naturalized citizen."[16]

Rents in Illinois had been $5 per acre from 1914 through 1917, then they had been increased to $6 in 1918. In the fall of that year, tenants were notified of a sharp increase—to $10 per acre for 1919. When trouble with tenants flared up, Fox explained to the press:

> We gave the farmers the advantage of four years of unusually high prices. They have made a lot of money. Increased income taxes and war's other calls on the Scully estate necessitated raising the rents this year. Too, we thought the war was going to continue. There is no doubt but that our farmers can make good returns under this new scale of rents.[17]

A compromise rate of $8 per acre, with no rents to be paid on land in legumes, ended a threat of tenant revolt in Logan County, where tenants had long tenure and large investments in improvements. This tenant unrest may have been nipped in the bud by the nonrenewal of a lease for a tenant of twenty years who was most active in opposing the rent raise. In words reminiscent of Irish tenant removal, he was "driven off the farm he . . . helped build by half a life-time of hard work . . . [by the] Scully Estate . . . a law unto itself." But events in Grundy County had gone too far. There, and in adjoining Livingston County, seventy Scully tenants formed a league to fight the new Scully demands. They refused either to pay rents or to move off the land. State Attorney Frank H. Hayes, of Morris, was retained by the association to carry the fight to the legislature and to enlist public sympathy in their cause. In the war of words between the association and the Scully estate, Hayes announced:

> We have arranged a meeting with the legislative agricultural committees to devise remedial legislation against such un-

American landlordism. We also have interested Gov. Frank Lowden, himself a practical farm man, in our case, and I believe we are assured of his support. Fortunately Illinois will have a constitutional convention in 1920, and we hope to embody in the new constitution some provisions against farm landlordism like that of the Scullys. We intend going about it on the ground of public policy. [To Hayes the Scully system] preserves the possession of the richest land in several communities in the hands of a few persons, a single estate, in fact. That, of course, limits the possibilities of our farmers by denying them the right to buy the land they desire and are able to buy. Moreover, it deprives a great many farmers of the liberty of action they should have.[18]

Cartoon from the *Prairie Farmer* at the time of post–World War I troubles between the landlord and his tenants.

All agreed that the Scully leases in Grundy and Livingston counties were specific in detail and legally unbreakable and that Scully, on the basis of existing law, could evict his defaulting tenants. Fox responded to Hayes and the association by saying:

We're going to give the farmers some more time. If they continue their fight we may turn the farms over to returned soldiers.

> We may have to help the latter at the start, but that can be arranged. Of course we shall lower rents if prices for farm products fall. We have told all the tenants this as well as the reasons for this year's increase.[19]

During this imbroglio, Frederick Trapp tried to soothe irritated feelings by taking some of the oldest tenants to Springfield to talk to the legislative committees.[20] He also described the general policy for the Scully estates by saying:

> A good farmer can make money under the Scully system. Mr. Scully always dealt fairly with his tenants. He wanted to make money and wanted them to make money also. In the early days he carried many of them along in bad years. He gave many of his tenants a start in this way. His farms are fertile and are kept so by the rotation of crops and by the adaptation of the latest approved methods of cultivation. We want only good farmers, industrious, thrifty men who pay their debts, want to prosper and do not drink. That is why we make the leases only for one year at a time—so we can get rid of the bad farmer in a short time. We don't want him burdening the soil any longer than that.[21]

In July, after hours of conferences and negotiations, the tenants who had been holding out for a return to the $6 per acre rent accepted the $8 compromise figure that had been used in Logan County. Also, it was further agreed that tenant improvements would be purchased by the Scully estate by July 15, 1919, and that tenants could borrow money from the estate for 5 percent rather than the 7 percent charged earlier.[22]

In Kansas, Nebraska, and Missouri, where per acre rents were $3.75 to $4.25, $2.40 to $4.50, and $2.50 respectively, tenant grumbling was also heard. Governors Henry J. Allen of Kansas and Samuel R. McKelvie of Nebraska called attention to the Scully estates as flagrant examples of tenant farming for which a legislative remedy was needed. One of the problems in Kansas in 1919 was the foundering of a plan to build a highway along the route of the Santa Fe Trail. Kansas highways of that era were paid for from taxes on adjoining property, and Scully tenants in Marion County refused to tax themselves for the proposed Santa Fe Trail highway, which would enhance their landlord's property.[23] It was not until the late summer of 1921 that tenant trouble, which newspapers termed a "revolt," erupted in Kansas.

Rental charges for Kansas land had been increasing for years, and the 1922 rates provided for no decrease. Since the price of farm products had declined to almost a prewar level, the tenants felt that they should get a proportionate decrease in cash rent. In Marion County the tenants developed a united organization, asked for a 40 percent reduction in rents, and sought abatement for the previous year because of poor crops. Some of the 350-member Scully tenant association made plans for leasing other land, and many made little effort to prepare their ground for the next year's crops. No lease renewals for 1922 were signed. Numerous meetings were held in rural schoolhouses and in the small communities of Antelope, Tampa, and Durham; and the tenant organization contended that they were facing a one-man lease that supplanted state law by requiring the tenant to waive his legal rights. Threats were made to strike against Scully prior to the wheat-sowing season if the landlord failed to provide relief.[24]

John Powers, the Marion County Scully agent, refused to budge on the terms of the 1922 lease; so, much that came later was a personality clash between Powers and the tenants. Powers wrote the Scully headquarters in Lincoln that the "Scully Tenant Union . . . has for its chairman not one of our tenants but a genuine Non-partisan leaguer, and for its Secretary a Socialist who is one of our oldest tenants and one up to this time has never given us any trouble." Powers was certain that Governor Allen was responsible for anti-Scully newspaper stories. He indicated his concern for the local agitation when he wrote: "At the meeting held in our office, Mr. [J. P.] Fengel, the Non-Partisan Leaguer did most of the talking. It seemed to me that he had his speech memorized as he told what the Legislature would do and also the organization of the Scully tenants of which he is the president, if their demands were ignored."[25]

The tenant union's organizational meeting on September 18 at Tampa had elected Fengel as president. A committee was appointed to draft a constitution and to report to the next meeting, eight days later. A survey of incomes from Scully farms in 1921 showed that Marion County tenants had "an average income of $1,030 per quarter section." Because the posted 1922 rental would be $600, plus taxes of a hundred dollars or more, they felt justified in asking for a 40 percent reduction. This would produce a rent of $360 per quarter section, which was still above the typical one-third collected by share-rent landlords in their area. The Scully agent offered a 25 percent reduction, but the Scully union of tenants held fast and finally settled for their initial demand, a 40 percent reduction for 1922. Most tenants signed their leases, but J. B. Shields, who lived on his own land and rented adjoining Scully land, did not get his lease renewed. He was secretary of the tenant association and had been a Scully tenant since 1883. In an argument with John Powers and Frederick Scully over the rents, Shields, in an aside, mentioned that Scully had just come from England, which had been saved by American boys in World War I. Scully left in an angry mood and evicted the longtime tenant. Word spread that Shields was to lose his lease; so, other Scully tenants and some other farmers quickly hurried to Marion to rent that land, an indication that the tenant association had backsliders.

Meanwhile, trouble of a different sort confronted Frederick Scully in Marion County. Godfrey Berg purchased his father's improvements in 1920 and began farming as a Scully tenant. Yearly leases were renewed until July 16, 1923, when Berg received notice "that his lease would not be renewed; that he must vacate by March 1, 1924; that he must dispose of his improvements on the leased premises after his rents had been paid; that he should not put in fall crops." Berg could find no one to purchase his improvements, and the agent made no effort to assist him. So, with the expiration of his tenancy, he did not immediately vacate the premises. The Scully agent got the county sheriff to move Berg's household goods and other personal property out onto the adjoining road.[26]

Powers reported to the Scully headquarters that Berg's land

> will be looked after this year by one of our tenants for a share of the crop. Mr. Scully, understands quite well the condition that prevails in that neighborhood among the tenants on account of the action we had to take to get possession of our land. Ralph and Charlie had to get some keys and go up and unlock the house before the tenant would go near the place, as there was a written notice on the door, "Keep Out Danger." Our purpose is to have the wheat and oats fully insured as soon as it is cut and in shock, if it is possible to do so.[27]

The tenant association was still active, and its members were anxious to test the legality of a Scully lease where nonpayment of rent was not an issue. Berg, an untypical, outspoken tenant, was willing to provide the basis for the litigation, and the association provided the money for legal fees. Berg filed suit against Scully to recover the value of his improvements. Tenant sympathy was overwhelmingly for Berg, in spite of his idiosyncrasies. When the case came up in the county district court, Berg's attorney sought to show that Powers lacked authority to sue or be sued in Scully's name. To their surprise he produced the necessary authorization. The customary transfers of improvements from one tenant to another were discussed, but Scully's attorneys contended that Berg had abandoned his property, which was thereby forfeited to the landlord. Judgment was in Scully's favor, and Berg received no satisfaction.

The Berg case was appealed to the state supreme court on the grounds that the lower court had erred by excluding evidence in Scully's "practice in dealing with his tenants." The court admitted the new evidence, which showed that an outgoing tenant would negotiate with an incoming tenant for the sale of his improvements. No new tenant was sent to purchase Berg's improvements, and because he had no crops in the ground, there was no prospective tenant interested in the lease. Even the idea of moving his improvements was considered unrealistic to the court, since the cost of dismantling and moving the property would almost equal its value. Seven of the nine justices of the Kansas Supreme Court sided with Berg, because "no provision of the contract contemplates forfeiture of the tenant's improvements to the landlord." The court mentioned a Kansas statute of 1925, which could not be retroactive for this case, but it did indicate the thrust of public policy, which "virtually places defendant's [Scully's] attitude . . . under public condemnation."[28]

By the 1920s both Thomas and Frederick Scully had married. Thomas had married Isabel Burrell but had divorced her after she had borne him four children—Sylvia, Iris, Sheila, and Douglas.[29] He had been married again, in 1924, to Violet Mary Simpson, of London. They had two sons, Michael John Scully, born in 1926, and Peter Dennys Scully, born in 1928. Frederick Scully had married Betty Gwendolyn Drake. They also had two sons—William Scully, born in 1922, and Robin Frederick Scully, born in 1924.[30]

E. Angela Scully continued to live in the house at 12 Holland Park. She divided most of her bonds, stocks, and securities among her three children on December 12, 1927. The greater part of her remaining property consisted of $2 million in United States Liberty loan bonds and her house in Washington.[31] She died in her London house on February 6, 1932, of "cardiac debility" and "chronic cirrhosis of the liver."[32] Her primary bequests were to her three children. Provision was made for her servants, five of whom were named in the will, with the largest amount, $10,000, to her nurse "if with her at time of death." Other servants with three or more years of service were to get one year's wages plus whatever was due to them. She also gave Frederick Trapp a sum of $30,000 and J. Edmond C. Fisher, the Beatrice agent, $15,000 "in memory of son William John Chynoweth Scully."[33]

No estate or death taxes had been paid when William Scully died in 1906, although the state of Nebraska had made an effort to collect an inheritance tax.[34] When Mrs. Scully transferred the land to Thomas and Frederick in 1918, the state's attorney for Logan County, E. Everett Smith, said: "Old Man Scully managed to dodge the Illinois inheritance tax before he died by transferring his holdings. But Mrs. Scully is not going to be successful in anything of that sort. We've got a suspicion that these recent transfers were made with just that purpose in mind, and we're taking steps to guard against any trick of that sort."[35] After Mrs. Scully's death, a fairly sizable tax was paid on the $2 million in bonds that her estate contained, but the federal government assessed an additional $3 million as an estate-tax deficiency on the assumption that the gifts of 1918 and 1927 had been made "in contemplation of death." One minor contention by the government was that Mrs. Scully, while she was an American citizen, had gone to England with no intention of returning to the United States; thus they contended that the normal $100,000 deduction for her estate should not be allowed.[36] For several years the Scully agents, their attorneys, and Thomas and Frederick Scully accumulated data, mostly gathered in England, to support their argument that the tax had been fully met. On April 1, 1936, the Board of Tax Appeals in Washington ruled in favor of the Scullys.[37]

Scully ownership of land in Bates County, Missouri, in the 1890s never settled down to be a comfortable alliance for the development of agriculture. Missourians adopted their "wait and see" attitude after initial belligerence, and by 1901, merchants in Butler made friendly comments about the enterprise of the 275 tenants on Scully lands. Compared to the cost of that land, income from it was far below that from any of the other Scully possessions. The uneasiness of the Scullys over the Bates County project is seen in correspondence between the Lincoln office and the Butler agents. Finally, after many years, Thomas Scully decided to dispose of that land in 1941 to the Farm Security Administration, which sold it to Missouri farmers who had been displaced by nearby army or war projects. Thomas was paid $1,078,150.53 for that land, which was not much different from the price paid for it by William Scully in the mid nineties, if one remembers that the elder Scully had immediately sold all improvements on the land.[38] Once the sale had taken place, Thomas Scully reportedly said that he had gotten rid of it because "you can never teach Missourians how to farm"; while local Bates County citizens responded in similar fashion, saying that it was "one of the best things that ever happened to Bates County" or that it was good riddance for "the antique and cold-blooded type of ownership that formerly existed."[39]

Over the years, many different agents served the owners of the Scully estates. Frederick Trapp retired in 1932, soon after the death of Mrs. Scully. His place was taken by his son, Will; and Frank Ryan became the new partner in the firm of Trapp and Ryan. In other agencies, sons or sons-in-law frequently followed in their elder's positions; but generally fewer personnel than the sixteen employed by William Scully in the early part of the century were needed to handle the vast estates. The last major change of Scully agents came in the late 1940s and early 1950s, when James M. Stewart, a former federal bank examiner, became the agent at Lincoln; J. M. Quackenbush headed the Beatrice office; and D. W. Montgomery was placed in charge in Marion. Each had a staff consisting of one assistant and one secretary, although occasionally part-time help was employed.[40]

The ups and downs of the 1920s and 1930s were behind them. National farm journals began noting the Scully estates in a favorable manner in the 1920s. During the 1930s there was recognition that Scully tenants had quickly adopted federal programs of the Agricultural Adjustment Administration, although there were times when the landlords abated part of the rent because of bad crop years. After many years of discussion about the harshness of the Scully lease, in 1933 the Kansas legislature passed a law making it illegal to use leases such as the one Scully had had his tenants sign. Although Scully was not mentioned by name, the explanatory paragraph preceding the law practically recited the Scully lease word for word. This law permitted a lien for no more than the "total crops grown on the leased land" and the "total receipts or returns from pasture."[41] Some changes were made in leases after 1933 because of this Kansas law.

During the 1940s and 1950s, commentary about the Scully estates in news media and agricultural journals noted the special care used to maintain soil fertility, to protect against erosion, and to provide for excellence in farming. Superlatives were frequently employed in describing the vast territory that the estates encompassed, the amazing prosperity of the tenants and the landlords, and the longevity of the entire operation.[42]

A detailed study of the Scully system in Marion County was made by agricultural economist Russell L. Berry during the period from 1947 through 1965. Berry interviewed large numbers of Scully tenants in Marion County and neighboring farmers. He found that the Scully system was comparable to the farming systems of English landed estates before mid-twentieth-century tenancy legislation. He described the Scully system thus:

> The rents are paid in cash, administered by professional agents or managers, tend to be raised slowly and only after yields and prices have fully justified the increase, are frequently lowered in years when poor yields or prices make rent payments difficult.[43]

Berry concluded that Scully rents in Marion County between 1947 and 1965 "appeared to be about half the cash rents charged by other landlords in Kansas."[44] Tenant improvements on these lands were often new, expensive homes and other needed outbuildings, but generally were not as high in quality as those on owner-operated farms. During interviews with Berry, some Scully tenants said that they had 99-year leases, thus they could justify the new improvements. When shown the one-year term in their contract, they modified their statement to: "Its just a community saying."[45] Thus, because of the relatively low cash rentals, a tenant-right in the improvements created an unusually high price for those improvements when a lease was transferred. Normally, that tenant-right in the improvements created no problem. By the mid twentieth century it was the practice of the

Scully estate agent in Kansas to offer vacated leases first to other Scully tenants.

But major problems occurred if the land were either condemned or sold. Thus, in the 1960s, when Congress authorized the Marion Reservoir in the Cottonwood Valley west of Marion, the federal government acquired "fee-simple ownership" of 12,500 acres, including slightly more than 3,400 acres of Scully land that was under lease to thirty tenants. The Corps of Army Engineers had never been confronted by such a system before, and it attempted to "purchase the land, including the improvements, from the landowner—leaving it to the landowner to settle with his tenants." The Scully estate refused the offer, claiming that it had no right to sell the tenants' improvements and that it was not under any "legal obligation to share the proceeds from the sale of the land other than on the basis of the fair market value of the improvements, which is much less than the amount paid for them by most tenants." A bill was introduced into Congress to direct the secretary of the army to acknowledge the tenant leasehold interest in the Scully land, and a tenant-appointed attorney appeared before the Congressional Committee on Public Works to explain the problem. Even the United States District Court of Kansas, on request of the tenants, issued special instructions to the land commission that was created to determine proper compensation for the land. Scully tenants were eventually "compensated for the full market value of their leasehold interests and as a result the Scully Estate . . . [received] less than the full value of the land itself." In customary usage the Scully lease gained a value over and above the value of the improvements involved.[46]

A change in ownership of part of the Scully estates came with the death of Frederick Scully at the age of sixty-three in 1942. He was on his way from Washington, D.C., to Lincoln, Illinois, when he had a heart attack in a Chicago hotel and died on October 28.[47] Frederick's wife, from whom he was estranged, had died as a result of a bicycle accident in England just six weeks earlier. She had been provided with a bequest of $100,000 and 400 acres in Marshall County, 1,521.77 acres in Gage County, and 6,667.40 acres in Nuckolls County, Nebraska. Their sons—William, aged twenty, a student at Northwestern University, and Robin, aged eighteen, a student at Harvard University—were the beneficiaries of both these estates. Frederick also made cash bequests to relatives, agents, and servants. The bulk of his estate was divided so that William received the Kansas land and Robin got the Nebraska land, with each getting a share of the Louisiana land.

Settlement of Frederick Scully's estate was delayed by the war and the size and complexity of his holdings. The probate court provided ancillary settlement on January 10, 1947, for the Kansas land and at about the same time in other states. Something over $4 million in estate and inheritance taxes was paid. Frederick, who had given each son a large sum of money earlier, stipulated in his will the income that each son should have at the ages of twenty-one, twenty-five, and thirty. He further directed that his sons would assume full management at the age of thirty-five, unless their guardian felt that they should have full possession at the age of thirty.[48] The trustee for William's part of the estate was John C. Scully of Peoria, Illinois; and William received his rights of full management at the age of thirty, in 1952. He had previously married and had established his residence at Beatrice, Nebraska, which became the headquarters for the Scully estates in Kansas and Nebraska.

The trustee for Robin Scully, inheritor of the Nebraska lands, was his uncle Thomas Scully. When Robin reached thirty years of age, his trustee, exercising his discretionary powers, refused to permit the owner to take over full management responsibilities on the grounds that he did not believe that Robin would carry out the Scully tradition. Robin took his case to the Nebraska district court in January, 1955, and its ruling favored his claim to his estate.[49] However, Thomas Scully appealed to the Nebraska Supreme Court, and Robin Scully's assumption of control of the Nebraska lands was thus delayed until 1959, when he was thirty-five years of age.

In the meantime, Thomas Scully was initiating an extensive reorganization of the ownership of his lands in Illinois. Not only was he concerned about high death taxes; income taxes were also high. Progressive rates for these taxes imposed heavier rates on large estates or large incomes, whereas the same estates and incomes, when divided into many hands, would not require the payment of as much total tax.

On May 12, 1954, Thomas gave each of his two sons about one-fourth of his land in a "trust for their lifetime and then to their children."[50] Michael, married and a graduate of Harvard, was given the land to the south of Logan County, and he established his home near Buffalo. Peter, married and a graduate of Princeton, got the Grundy and Livingston County estates, and he located his home on some of his land near Dwight. By 1959 Thomas Scully had numerous grandchildren, so he established a trust for them with about twelve thousand acres of Logan County land. Provisions of that trust would permit future grandchildren to share in it, and all property of the grandchildren would go to their children, who could sell or trade the land if they so wished. But "they must first offer it to a male descendant of Thomas Scully by the name of Scully."[51]

Both Thomas and Frederick Scully owned houses in France, near Cannes. Frederick generally regarded Washington, D.C., as his home, although he spent much time in England and France. Thomas and his family resided at their home, "Bastide," until World War II drove them from France. Later, Thomas said that he must have crossed the Atlantic a hundred times in his lifetime. In 1941 he and his wife built a large fine home on an eighty acre tract of Scully land near Lincoln, Illinois. That remained his home for the remainder of his long life, and Thomas Scully thus became better known to his tenants than did his father. He still did not associate with tenants until his last few years, when he found that he enjoyed visiting with some of the oldest tenants and that they enjoyed him. After the war, Thomas and his wife again spent long periods of time in their French home, where it was learned in the late 1950s that he had Parkinson's disease. On July 12, 1961, he died in a Chicago hospital. He had been under treatment for chronic pneumonia for more than a year.[52]

Thomas Scully's will contained the following passage, suggesting that he had learned his father's prescription for the land very well:

> I authorize and strongly recommend the continuance of the practises developed by my father, my brother and myself. . . . I authorize and strongly recommend the retention of all of the land which I shall own at the time of my death.[53]

His bequests gave half of his remaining twelve thousand acres to his wife under a marital trust and, on her death, to her sons; the other half was "placed in a residuary trust for the use of his wife during her lifetime and then to her sons."[54] Upon probating, his estate was given a value of $3 million in personal property and $3 million in real estate.[55] Before he began to divide his Illinois property among his sons and grandchildren, Thomas Scully's property probably had a value of five times that amount.

By 1964 the cash rent on Scully land in Illinois averaged $23.51 per acre, which included taxes. One-fourth of the cultivated land was seeded to legumes or a mixture of legumes and grass, for which no rent was collected. A four-year rotation, established in 1926, provided for corn, corn, and oats in sequence, with a catch crop such as clover seeded in the oats. The lease permitted the substituting of soybeans for corn or of wheat for oats. This rotation provided a low corn base on Scully land, which was a handicap in governmental programs existing in the 1960s. So, in 1975, the phraseology and content changes in the Illinois lease moved entirely to a corn and soybean rotation.[56] In Grundy and Livingston counties the Scully leases were dominantly crop-share in format, which they had been since the troubles after World War I. There were twenty-seven crop-share, seven cash-rent, and one corporation lease for that area, and the average size of a farm was 325 acres.[57]

Also in 1964 the average cash rent on Scully land in Marion County, Kansas, at $3.98 per acre, was more than double the 1947 rate, and taxes were listed separately as $2.02 per acre.[58] Allowances were made for land in legumes or land in summer fallow. Leases were much briefer than those in Illinois and did not specify a rotation plan for the land. Marion County was in small-grain country, and livestock production was an important element in the area's agriculture. Encouragement was also provided for other soil-conserving procedures, such as spreading agricultural lime or trace minerals.

In all areas the numbers of tenants on the Scully lands declined drastically in the years after World War II. Tenants were generally given preference for assuming vacated leases, although some agents still permitted lessees to find their own successors. In Illinois in 1965, "modestly improved 160-acre leases [sold] for about $25,000"—that is, an incoming tenant would pay an outgoing tenant that much for his improvements. It was not uncommon in the same period to find leases in Kansas selling for as much as $100 per acre, and all the improvements that an outgoing tenant was providing in some cases were the fences.[59]

By the mid 1960s there were 175 farm units on Scully land in Logan and Sangamon counties. These leases were valued, and "usually not more than five leases change hands annually. . . . generally due to retirement of an aged tenant, and then the lease is often turned over to a son."[60] A similar situation was found in Marion and Dickinson counties in 1977—165 tenants and about five changes each year.[61] The total numbers of tenants in Illinois, Kansas, and Nebraska have declined to something under 600. Acreage in Scully ownership has also declined since 1906, when William Scully had amassed almost 225,000 acres of American land in four states. William gave the children of John Scully slightly more than 9,000 acres, and slight reductions came through the sale of small tracts for public use. In the 1920s Frederick Scully disposed of three isolated tracts in Marion County totaling 160 acres. In 1941 Thomas Scully sold all of the Missouri land, 41,844 acres. About that time the city of Hillsboro, Kansas, condemned more than a section of nearby Scully land for public use. In 1959 William Scully, Frederick's son, sold 3,320 acres of Nuckolls County land to tenants. He had gotten that land from his mother, who had died prior to the death of his father. The development of the Marion Reservoir in the mid 1960s cost William an additional 3,400 acres. Thus, the Scully heirs in 1977 still held nearly 175,000 acres of the accumulated land owned by William Scully. In addition, Frederick's purchase of Louisiana land during World War I raised the total to just over 200,000 acres.

Sale of land in the twentieth century has had far different consequences to the estates from sales by William Scully in the 1850s or even the 1880s. It became virtually impossible for the Scully Estates to replace these lands with new purchases while remaining on good terms with people in the community. Good tenants were always on the lookout for additional land, and the Scully family would have been severely criticized for competing in the land market with tenants. Thus, the inevitable decline in acres in the family estates has brought on more intensive use of the land. Two leading examples are the incorporation of a portion of Peter Scully's land with the assets of an outstanding tenant and the entrance of William Scully into cattle feeding in a big way.

Peter Scully and one of his tenants, Eugene Merrifield, developed a farm corporation in the early 1960s. Merrifield's Scully lease had grown to 1,100 acres, and he was fattening a thousand or more head of cattle each year. Through incorporation, Peter Scully put the land into the new corporation, while Merrifield supplied the management, implements, and livestock. Each owned half of the stock; and Scully served as the president and received his returns on stock earnings. Merrifield, as full-time manager for the corporation, received a salary and his stock dividends. Farm corporations in Illinois are few, but the advantages of the Scully-Merrifield partnership are considered a mutual benefit to the owners.[62]

About the same time, William Scully developed a ranch headquarters and started elaborate feeding operations on a section of Scully land located almost three miles west of Durham, Kansas. A cattle feed lot efficiently handles a large number of cattle each year on this tract. William Scully takes up residence at the ranch for part of each year. Thus, the owner of the Kansas land spends more time on his land than either his father or grandfather did. In 1977 his residence was in Portland, Maine, where he moved from Beatrice, Nebraska, many years ago.

Robin Scully, owner of the Nebraska land, maintains his official residence in Beatrice, Nebraska, but he spends much of each year at his house north of London or with his racing horses on some property not far from Golden in county Tipperary. Robin is the only one of William Scully's grandsons who has not married.

The death of Mrs. Violet Scully, in a village near her French country house in the late summer of 1976, ended the ownership of any of the Scully land by those in the second generation. Thus, all of the owners by 1979 were grandsons, their spouses, or their children. The great-granddaughters share in the estate with their brothers, so discrimination against the female members of the family has ended, and they have come into ownership of land through a variety of trust arrangements.

By the late 1970s most Scully tenants are the type of good farmer that was sought in a Scully lessee much earlier. They are efficient producers on crop and livestock farms and have gained considerable wealth. In Illinois, the tenants are required to live on the Scully lease, whereas, in other states, the tenants often live on land that they own, while they lease Scully land to enlarge their farm unit. Minor variations exist in these states as to how taxes are paid. Increasingly, the Illinois lessees were concentrating on crops, while in the western states the major income for lessees was in livestock enterprises.

The Scully estates of the 1960s and 1970s, under the ownership of a third generation of Scullys, has moved to accept the best available advice, in order to keep pace with rapid changes in agriculture. The small holdings retained in Ireland are of little economic value and are kept as a novelty.

Most of the Scully land in Kansas and Nebraska has been in the Scully family for more than a century, the Louisiana land has been owned for at least fifty years, and the Illinois property has been a family possession for one hundred and twenty-five years. Changes have come slowly to this mammoth estate, but the leadership has always readily accepted the steady long-run advantages over the more risky short-run gains. An apparent willingness to charge lower cash rents than those charged by neighboring landlords is evident, perhaps a practice that was developed to produce greater tenant support and appreciation for the Scully ownership. The primary owners of the Scully estates are now in their fifties, and the principal agents are some twenty years older and approaching retirement. No doubt there are plans for the future maintenance of the Scully estates, just as were made by the first William Scully. His shadow looms large over any action taken by his successors.

Bibliographical Essay

wswswswswswsws

The personal papers of William Scully, if there ever were any, were lost in his London home during World War II. Some of his business records and correspondence are retained in the offices of the Scully Estates in Lincoln, Illinois, and in Beatrice, Nebraska. Without them, the Scully biography would have been limited to newspaper stories (which are largely hostile), to interviews, and to public records that contain many entries which apply to William Scully.

Essential to this study were the private business records of William Scully found in his estate offices. Public records have been fully explored in the Logan County, Illinois, courthouse, the Marion County, Kansas, courthouse, and in the public land records of the old General Land Office, now located in the General Records Center, Suitland, Maryland. Some public records, appropriate to the Irish years, were found in the Office of State Papers, Dublin Castle, and important newspapers were located in the National Library, Dublin. Other Irish and English newspapers were found in the British Museum Newspaper Library, Colindale.

Significant additions to this biography were found in newspapers and manuscripts located at the Kansas State Historical Society, Topeka, at the Nebraska State Historical Society, Lincoln, and in the Black and Williams Papers as well as the Oglesby Papers of the Illinois State Historical Society, Springfield. Farrell Library, Kansas State University, had most of the general information needed, and the Case Collection, of the Spencer Research Library, University of Kansas, provided unique holdings.

Discussion of the life and activity of William Scully with tenants, agents, and owners, as well as with knowledgeable onlookers in the vicinity of the Scully lands, was an important source for the elusive story of Scully landlordism.

After I wrote my master's thesis on "The Scully Land System in Marion County, Kansas," which was completed at Kansas State University in 1947; Paul James Beaver prepared a master's thesis with the title "William Scully and the Scully Estates of Logan County, Illinois," which was completed at Illinois State University in 1964. Russell LaVerne Berry's doctoral dissertation in agricultural economics on

"The Scully Estate and Its Cash-Leasing System in the Midwest" was completed at Ohio State University in 1966. Each of these contains original materials not found elsewhere. Their combined bibliography, somewhat overlapping, covers twenty-one pages. For specific sources used in this biography see the citations in the notes.

Notes

wswswswswswsws

CHAPTER 1

1. Nenagh (Ireland) *Guardian*, August 15 and 19, 1868; London *Times*, August 22, 1868.
2. Cemetery at Shronell, 3½ miles west of Tipperary, Ireland. The text on the monument reads:

   Ballycohey
   1868–1968
   In Memory of
   The Fight Against Landlordism
   Michael O'Dwyer
   Michael Hanley
   Patrick Quinn
   William Quinn
   John Heffernan
   Timothy Heffernan
   John Ryan
   Denis Hayes
   Laurence Hayes
   John Greene
   John Hanrahan
   Kenneth Twomey
   Patrick Greene Michael Foley
   Who With Firearms Resisted Eviction
   on 14th August 1868, and So Frustrated the
   Despotism of Alien Landlords, with the Aid of
   The Men and Women of Ballycohey.
   Grant Them O Lord Eternal Rest.
3. John B. Burke, *Burke's Genealogical and Heraldic History of the Landed Gentry, Including American Families with British Ancestry* (London: Burke's Peerage, Ltd., 1939), p. 2020, gives the lineage of William Scully. With only

one exception, sources state that he was born in Kilfeacle; the exception says Dublin.

4. Patrick O'Farrell, *Ireland's English Question: Anglo-Irish Relations, 1534–1970* (New York: Schocken Books, 1971), p. 122.
5. Patrick Kelly, *Irish Family Names* (Chicago: O'Connor & Kelly, 1939), p. 116. Similar Irish names were O Scollee, O Scully, Skelly, Scallon, and Scally, all from the Gaelic *sgolai(d)he,* meaning schoolman. Synonyms were "a learner, a scholar, a schoolboy." See also Edward MacLysaght, *Irish Families* (Dublin: H. Figgis, 1957).
6. Burke, *Landed Gentry,* p. 2020.
7. Paul J. Flynn, *The Book of the Galtees and the Golden Vein* (Dublin: Hodges, Figgis & Co., 1926), pp. 63, 132–33, 181, 322.
8. Thomas Laffan, *Tipperary's Families: Being the Hearth Money Records for 1665–6–7* (Dublin: J. Duffy & Co., 1911), pp. 60, 123.
9. Thos. W. H. Fitzgerald, *Ireland and Her People,* 5 vols. (Chicago: Fitzgerald Book Co., 1909–11), 1:346–47. Some sources say that Denys was the second Catholic student at Cambridge. Fitzgerald states (5:742–43) that the greater part of the surface of county Tipperary "is level, and much of the soil is very fertile, especially in the Golden Vale, which is calcareous loam." Generally, the term "Golden Vein" is used for the area, especially to the northwest of Rock Cashel.
10. Manuscript book, "William Scully—Index to Deed Book made in /56," Office of the Scully Estates, Lincoln, Illinois. This contract was reputedly on record in the register office of the city of Dublin, dated December 10, 1808, in bk. 605, p. 389 and no. 412623.
11. Catherine was the daughter of Vincent Eyre of Highfield and Newholt, county Derby. Thus, William and his brothers and sisters were half English. A contract known as a marriage settlement, which involved various lands, was drawn up between specified members of Catherine's family and James and Denys Scully. It was signed September 7, 1808.
12. There is no author's name on the title page of this pamphlet. In the entry catalog at the National Library, Dublin, the title and other publication data are preceded by [D. Scully]. Much is made of the inequity of British law in Irish writings, and the example of imprisonment for the publisher, with no such penalty for the writer, is one of them.
13. Extract on stationery inscribed "70 Holland Park, London, W.—— 18—," in files of the Office of the Scully Estates, Lincoln, Illinois. This material, which is unpaged, is said to be a translation from the French of *Popular Scenes in Ireland* (Paris, 1830). Vincent D. Scully, of Montreal, Quebec, Canada, in a letter to the author, dated February 6, 1966, says that the defense costs in this family court suit were £30,000.
14. *Blackwood's Edinburgh Magazine* 15 (March, 1824): 285.
15. St. Louis *Post-Dispatch,* March 31, 1901.
16. *Tipperary Free Press* (Clonmel), November 30, 1842.
17. Molly O'C. Bianconi and S. J. Watson, *Bianconi: King of the Irish Roads* (Dublin: Allen Figgis, 1962), p. 116.
18. *Tipperary Free Press,* November 30, 1842; Dublin *Evening Mail,* November 30, 1842.
19. Dublin *Evening Post,* November 29, 1842.
20. Letter from John Macleod, Tipperary, April 26, 1842, to the Under Secretary in

Dublin Castle, in Tipperary County Outrage file, 1842, Office of State Papers, Dublin Castle. In Ireland, conacre referred to the subletting, for a single crop, of small portions of a farm that had previously been prepared for sowing or planting; so, rates were usually very high.

21. Macleod to Under Secretary, April 26, 1842, Tipperary County Outrage file, 1842.
22. London *Times*, October 21 and 28, 1842.
23. Macleod to Under Secretary, April 26, 1842, Tipperary County Outrage file, 1842.
24. Letters from Thos. Annom, Limerick, to E. Lucas, Esq., October 31 and November 4, 1842, Tipperary County Outrage file, 1842.
25. *Tipperary Free Press*, November 30, 1842; John Macleod to Under Secretary, November 27, 1842, Tipperary County Outrage file, 1842; London *Times*, December 2 and 6, 1842.
26. *Tipperary Free Press*, November 30, 1842; John Macleod to Under Secretary, November 27, 1842, Tipperary County Outrage file, 1842; London *Times*, December 2, 1842.
27. London *Times*, December 26, 1842.
28. Ibid., December 6, 1842.
29. Tipperary County Outrage file, 1842, Abstract Report of Outrage.
30. Letter to E. Lucas from John Roberts of Cashel, December 6, 1842, Tipperary County Outrage file, 1842.
31. Letter, Wm. Kemmis, Cr. Sol., to T. N. Redington, Esq., from Kildare St., August 18, 1843, Tipperary County Outrage file, 1843. The cost of the Scully murder case investigation was £131 .. 8 .. 11. It was quickly established that Thomas Murname, Scully's herdsman, was murdered by kinsmen of a man whom he had killed in 1838. He had been acquitted of that crime.
32. Andrew Coffey file, CSO RP, 27-169 1851, Office of State Papers, Dublin Castle.
33. London *Times*, December 28, 1847. Vincent Scully was, at that time, a member of Parliament.

CHAPTER 2

1. London *Times*, February 18, 1843.
2. Copy of the will of Denys Scully, dated August 3, 1830, from George M. Curtis, III; manuscript book, "William Scully—Index to Deed Book made in /56," Office of the Scully Estates, Lincoln, Ill., pp. 45–48. The family agreement was called a "Deed of Family Compromise."
3. Ibid., pp. 93–98.
4. "A/c of Decennial increase in value of my property—October 4th/79," notes in William Scully's handwriting in the Office of the Scully Estates, Lincoln, Ill.
5. Ibid.
6. A. M. Sullivan, *New Ireland: Political Sketches and Personal Reminiscences of Thirty Years in Irish Public Life*, 16th ed. (London: Burns Oates & Washbourne, Ltd., 1877, 1882), p. 365.
7. Manuscript book, "William Scully—Index to Deed Book made in /56," Office of the Scully Estates, Lincoln, Ill., p. 1. Such a lease would expire at the death

of the survivor of the two names on the lease, when it could then be renegotiated with the landlord.

8. "Clonmel 15 March 1833 Grand Jury Book (Criminal) to Summer Assize 1850," manuscript volume in the courthouse, Clonmel, Ireland. These criminal charges suggest that no one was hit in what must have been a wild affair.
9. St. Louis *Post-Dispatch*, March 31, 1901. From all available evidence, Scully believed that he came at the invitation of the United States government. Paul W. Gates suggests that the passage by Parliament of the Act of 1849, creating the encumbered estates court, enabled Scully to sell some of his Irish land, a thing that he did not do. See Paul Wallace Gates, *Frontier Landlords and Pioneer Tenants* (Ithaca, N.Y.: Cornell University Press, 1945), pp. 35–37, reprinted from *Journal of the Illinois State Historical Society*, June, 1945.
10. Ibid.
11. Interview with Leland Miller, Lincoln, Ill., November 19, 1963.
12. *History of Logan County, Illinois* (Chicago: Inter-state Publishing Co., 1886), p. 366.
13. St. Louis *Post-Dispatch*, March 31, 1901.
14. General Land Office Records, "Abstracts, M.B.L.W. Locations, Act of 1847, Illinois," vol. 36, National Archives. Scully made use of 54 of the 181 military land warrants used at Springfield in October, 1850.
15. Paul W. Gates, *History of Public Land Law Development* (Washington, D.C.: Government Printing Office, 1968), pp. 270–78.
16. "Some History of the Scully Family," manuscript in the Office of the Scully Estates, Lincoln, Ill.

CHAPTER 3

1. "Land-office business" was a popular nineteenth-century American term for hectic, demanding activity such as occurred in many federal land offices that were selling underpriced land.
2. Chicago *Morning News*, May 5, 1887.
3. Patents on file in the Office of the Scully Estates, Lincoln, Ill.
4. Paul James Beaver, *William Scully and the Scully Estates of Logan County, Illinois* (n.p., 1964), pp. 16–17. Beaver reports that the *Illinois State Journal*, located in Springfield, contained such advertisements on May 10, June 7, July 1, and September 14, 1850. Many years later, on October 19, 1859, the Lincoln *Weekly Herald* advertised: "Wanted.—A few land warrants.—For information apply at this office."
5. "Abstracts, M.B.L.W. Locations, Act of 1847, Illinois," vol. 36, National Archives. Lands adjacent to Scully's entries were entered as early as 1836.
6. Manuscript book, "William Scully — Index to Deed Book made in /56," Office of the Scully Estates, Lincoln, Ill.
7. "Abstracts, M.B.L.W. Locations, Act of 1847, Illinois," vol. 36, National Archives. Fifty of the fifty-five receipts from the Chicago Land Office for the Grundy County land had an attached statement showing the use of the "Military Bounty Land Act of 28 September, 1850"; however, a letter from Jane F. Smith, Office of the Civil Archives, National Archives, April 14, 1964, shows that all of the Scully locations of March 24, 1852, used the Military Warrants

of the 1847 Act rather than 1850. There is much related evidence to support the fact that the young married couple moved to Illinois in 1852.

8. Beaver, *William Scully,* p. 12, from Federal Survey Notes, bk. 126, pp. 55-60, 69, 80, 91; bk. 137, pp. 180, 241; bk. 138, pp. 124–46; September, 1823, Archives Building, Springfield, Ill.
9. John Mason Peck, *A Guide for Emigrants, Containing Sketches of Illinois, Missouri, and Adjacent Parts* (Boston: Lincoln & Edmands, 1831), p. 106, as cited in Margaret Beattie Bogue, "The Swamp Land Act and Wet Land Utilization in Illinois, 1850–1890," *Agricultural History* 25 (October, 1951): 171.
10. Beaver, *William Scully,* p. 13, from an 1844 map of Logan County in the law office of Miller and Miller, Lincoln, Ill.
11. Gates, *Frontier Landlords,* p. 36. Between 1852 and 1857, Scully's purchases in Logan County amounted to more than $5,000, with one 1856 purchase accounting for three-fourths of the cost and acreage.
12. Beaver, *William Scully,* p. 19.
13. "Address" of Thomas A. Scully to the annual meeting of the Logan County Historical Society at Emden, Ill., February 24, 1957. Copy in Office of the Scully Estates, Lincoln, Ill.
14. Allan G. Bogue, *From Prairie to Corn Belt: Farming on the Illinois and Iowa Prairies in the Nineteenth Century* (Chicago: University of Chicago Press, 1963), p. 84.
15. According to a certified copy of the application in the Office of the Scully Estates, Lincoln, Ill.
16. This map, whose dimensions are 31″ x 45″, may be found in the Office of the Scully Estates, Lincoln, Ill.
17. Genealogy of the Scully Family, prepared by Vincent Scully, Montreal, Canada, in the Library of William Minogue, caretaker of Rock Cashel, Cashel, Ireland. Since Gertrude is called Mary in William's will, she must have had both names. Both Gertrude and Julia were later married. Kathleen, in later years, received grants of money from her father, and he requested in his will that they be continued.
18. Beaver, *William Scully,* pp. 19–20. A John Hickey, who was living "on what is known as the Skully *[sic]* farm," advertised in the Lincoln *Weekly Herald,* April 13, 1859, that he had found a stray horse.
19. *Illinois State Journal,* August 31, 1855.
20. "Calcul'n — sale of land to Straut [which may have been Jacob Strawn] or others made May 1/55," manuscript in William Scully's handwriting in the Office of the Scully Estates. Scully figured disposal of his property as follows:

    1855 — 10 tracts of 160 acres each at $5 per acre
    1856 — 20 tracts of 160 acres each at $5 per acre
    1857 — 20 tracts of 160 acres each at $6 per acre
    1858 — 20 tracts of 160 acres each at $7 per acre
    1859 — 40 tracts of 160 acres each at $8 per acre
    1860–1 — 65 tracts of 160 acres each at $10 per acre
21. Margaret Beattie Bogue, *Patterns from the Sod: Land Use and Tenure in the Grand Prairie, 1850–1900* (Springfield: Illinois State Historical Library, 1959), p. 87.
22. Two sales—one for eighty acres in 1856 and one for forty acres in 1857—are noted in official county records, but not in the "Tax book." Scully realized a

nice profit on incompleted land sales, although court action was frequently necessary.

23. "Miscellaneous Record," vol. 1, p. 227; vol. 24, p. 442, Office of the Logan County Recorder, Lincoln, Ill.
24. "Tax book," vol. 7, pp. 188, 516, Office of the Logan County Recorder, Lincoln, Ill.
25. "Miscellaneous Record," vol. D, p. 6; vol. V, p. 265; "Mortgage Record," vol. D, p. 60, Office of the Logan County Recorder, Lincoln, Ill.
26. Letter from Parks to John Williams and Company, March 3, 1860, Black and Williams Papers, Illinois State Historical Society.
27. Paul W. Gates, "Land Policy and Tenancy in the Prairie States," *Journal of Economic History* 1 (1941): 76.
28. File 312, record vol. 2, pp. 191, 387, Logan County Court, Lincoln, Ill.
29. Files 41, 173, and 193, record vol. 1, pp. 113, 130, 249, 270, 404, 567, and 578, Logan County Court, Lincoln, Ill.
30. Contract of July 7, 1856, with Ira Martin, Sr., specified John Williams of Springfield. Several 1855 contracts said that payment would be made in Springfield, and some said Chicago.
31. Beaver, *William Scully*, p. 23.
32. Parks to Williams, April 20, 1860, Black and Williams Papers. In spite of this arrangement, there were delinquent taxes in later years.
33. Parks and McGalliard were leaders of the Logan County Republican party.
34. Marginal notes on patents, Office of the Scully Estates, Lincoln, Ill.; letter from Scully to John Williams and Company, August 10, 1859, Black and Williams Papers.
35. Parks to John Williams and Company, March 3, 1860, Black and Williams Papers; deposition in case of *William Scully* vs. *John McMullen*, filed August 8, 1857, Office of the Logan County Recorder, Lincoln, Ill.
36. "Miscellaneous Record," vol. 1, pp. 209–11, Office of the Logan County Recorder, Lincoln, Ill. The names of the tenants were Woolrod Keif, Charles Wehner, Michael Hoffman, Ernest Leirer, and Daniel Behler.
37. Letters, William Scully to John Williams, 1860, Black and Williams Papers.
38. Ibid., May 28, 1861.
39. Ibid., May 14, June 16 and 17, 1863.
40. Ibid., June 22, 1865.

CHAPTER 4

1. P. S. O'Hegarty, *A History of Ireland under the Union, 1801 to 1822* (London: Methuen & Co., 1952), p. 462.
2. E. D. Steele, *Irish Land and British Politics: Tenant-Right and Nationality, 1865–1870* (London: Cambridge University Press, 1974), pp. 3–25.
3. London *Times*, June 6, 1870. As a queen's counsel, Vincent could place the initials "Q.C." after his name.
4. *Dictionary of National Biography* (London: Oxford University Press, 1968), vol. 17, p. 1097. When Vincent died on June 4, 1871, his residence was at St. Johns Wood, London.
5. London *Times*, December 28, 1847.

6. Ibid., August 26, 1857. Vincent Scully's *net* annual rental in 1857 was said to have been £2,100.
7. Genealogy of the Scully Family.
8. Ibid.; London *Times,* April 14, 1857. The election contest probably refers to the race of his brother Vincent for a parliamentary seat.
9. Genealogy of the Scully Family.
10. Deeds on file at the Office of the Logan County Recorder.
11. "A/c of the Decennial increase in the value of my property—Oct. 4, 1879," Office of the Scully Estates, Lincoln, Ill.
12. Scully letters in Black and Williams Papers.
13. Kilkenny *Journal,* August 2, 1865.
14. Ibid.
15. Ibid., June 7, 1865; Cashel *Gazette and Weekly Advertiser,* June 10, 1865.
16. Kilkenny *Journal,* June 7, 1865.
17. Ibid., August 2, 1865.
18. Ibid.
19. A special jury was discussed in a report from the *Irishman* in the Kilkenny *Journal* of August 16, 1865. The *Irishman* said that the case had a very unusual outcome, considering the fact that "the jury was *a landlord* jury—a special jury, of the twelve, ten are J.P.'s."
20. Kilkenny *Journal,* August 2, 1865; London *Times,* August 5, 1865.
21. Kilkenny *Journal,* August 5, 1865; London *Times,* August 5, 1865.
22. Kilkenny *Journal,* August 5, 1865. Three prisoners, sentenced just after Scully, received three months' imprisonment each, with hard labor. Their crimes were gaol-breaking, for two of them, and "assault upon a man . . . whose death took place in a short time afterwards, . . . not occasioned by the assault," for the other.
23. Beaver, *William Scully,* p. 30.
24. Kilkenny *Journal,* August 5, 1865. Scully made no mention of these cases in his letters to John Williams. In a letter of January 23, 1865, to John Williams, well before Gurtnagap activities came up, Scully wrote, "I do not see much good that I can do in Illinois this year, & I reckon not going until 1866—," Black and Williams Papers.
25. Kilkenny *Moderator,* August 5, 1865.
26. Kilkenny *Journal,* August 9, 1865. The *Journal* expressed surprise that Scully was convicted in the criminal suit.
27. Ibid., August 23, 1865, from the London *Weekly Register.*
28. Sullivan, *New Ireland,* p. 365.
29. Patrick O'Donovan, "The Battle of Ballycohey—perfect example of a just war," undated clipping from an August, 1968, Irish newspaper; Steele, *Irish Land,* p. 71.
30. Sullivan, *New Ireland,* p. 366.
31. St. Louis *Post-Dispatch,* March 31, 1901; remarks of T. A. Scully to J. M. Stewart, June 9, 1958, Office of the Scully Estates, Lincoln, Ill.
32. Michael Davitt, *The Fall of Feudalism in Ireland: or, The Story of the Land League Revolution* (London: Harper, 1904), p. 76.
33. Godfrey T. Locker Lampson, *A Consideration of the State of Ireland in the Nineteenth Century* (London: Archibald Constable & Co., 1907), p. 337; Sullivan, *New Ireland,* p. 365; "The Centenary of Ball[y]cohey," *Tipperary Star* (Clonmel, Ireland), August 17, 1968.

34. Copy of lease, Office of the Scully Estates, Lincoln, Ill.
35. Sullivan, *New Ireland*, p. 364; London *Times*, August 18, 1868.
36. London *Times*, September 14, 1868.
37. Sullivan, *New Ireland*, pp. 369–70.
38. Nenagh *Guardian*, August 19, 1868.
39. Sullivan, *New Ireland*, p. 366; according to Patrick O'Farrell, *England and Ireland since 1800* (London: Oxford University Press, 1975), pp. 167–68, Irish agrarian relations in the 1850s resembled guerrilla warfare. Landlords and their agents were always armed, and peasants killed them if they could. One landlord in 1882 used an armed force of one hundred to collect rent that was in arrears from one of his tenants.
40. Nenagh *Guardian*, August 15, 1868.
41. Sullivan, *New Ireland*, p. 367; Nenagh *Guardian*, August 15, 1868.
42. Sullivan's chapter on "Ballycohey" in *New Ireland* is the most popular source of this Agrarian Outrage. Extensive coverage appeared in the London *Times* on August 15, 17, 18, 19, 20, 21, 25, 31, September 4, 7, 11, 14, October 8, 14, and November 1, 1868. The Nenagh *Guardian*, August 15–19, 1868, and other Irish newspapers and published histories, such as Davitt, *The Fall of Feudalism in Ireland*, pp. 76–77; and O'Hegarty, *A History of Ireland*, pp. 462–65; and James Godkin, *The Land-War in Ireland* (London: Macmillan & Co., 1870), p. 210, all report this incident in varying degrees. Quotations that follow are drawn from a combination of these sources in an effort to present the narrative as completely and accurately as possible.
43. *Tipperary Star*, August 17, 1968; letter to the editor from D. V. Kent, Kansas City *Star*, February 2, 1919; Thomas McEnery, monument builder at Shronell Cemetery, to author, October 26, 1971.
44. Michael Francis O'Dwyer, *The O'Dwyers of Kilnamanagh: The History of an Irish Sept* (London: John Murray, 1933), p. 321.
45. Those who were arrested included John Greene and John and Timothy Heffernan, who were reputed to have been among the Ballycohey defenders.
46. Dublin Castle, Office of State Papers, "File on Ballycohey Outrage," letter of September 9, 1868, from T. A. Larcom, Dublin Castle, to the magistrates at petty sessions, Tipperary, which states that reinforcements would be stationed at Glenbane, Monard, Mooresfort, and Tipperary. The Clonmel *Chronicle*, February 13, 1869, reported that the cost of the eight extra police for forty-five days was assessed against the townland of Ballycohey.
47. Nenagh *Guardian*, August 19, 1868.
48. Ibid.
49. Ibid.
50. Ibid.; London *Times*, August 21, 22, and 25, 1868.
51. London *Times*, August 31, 1868.
52. Remarks of T. A. Scully, June 9, 1958, to J. M. Scully, and copy of a letter from Mrs. Ita Parry, of London, to Mrs. T. A. Scully, July 3, 1962, in Office of the Scully Estates, Lincoln, Ill.
53. Sullivan, *New Ireland*, p. 368.
54. "Some History of the Scully Family," Office of the Scully Estates, Lincoln, Ill.
55. London *Times*, August 21, 1868.
56. Ibid., September 7, 1868.
57. Dublin Castle, Office of State Papers, "File on Ballycohey Outrage."
58. Sullivan, *New Ireland*, p. 373.

59. London *Times,* October 8 and 14, 1868.
60. Roger Chauviré, *A Short History of Ireland,* trans. by the Earl of Wicklow (New York: Devin-Adair, 1956), p. 109. O'Dwyer, *The O'Dwyers of Kilnamanagh,* pp. 324–25, and Steele, *Irish Land,* pp. 65–90, have much on Gladstone's acquaintance with the Ballycohey incident. On p. 73, Steele quotes a campaign speech by Gladstone in Liverpool on October 14, 1868, when he said that "because Scully's action had been perfectly legal 'everyone of us . . . cannot indeed justify, but can excuse, or if we cannot excuse, can at least understand . . . this deep and sullen feeling of . . . passive estrangement, something arising into active and burning hatred,'" which confronted Britain because of Irish land laws.
61. *Weekly Illinois Courier* (Lincoln), October 31, 1906, as found in Beaver, *William Scully,* p. 33. This ballad was the work of the Fenian Charles Joseph Kirkham (1830–82). Slievenamon Mountain, elevation 2,368 feet, is located 15 miles southeast of Cashel.
62. *Tipperary Star,* August 17, 1968; underlining provided in order to show Scully connection.
63. Ballycohey monument in Shronell Cemetery, located 3½ miles west of Tipperary and near Ballycohey.

CHAPTER 5

1. "Net income from American lands, after paying for improvements on Ill^s lands, & after paying Neb & Kansas taxes—," notes in William Scully's handwriting, Office of the Scully Estates, Lincoln, Ill.
2. "A/c of the Decennial increase in value of my property—Oct. 4, 1879," notes in William Scully's handwriting, Office of the Scully Estates, Lincoln, Ill.
3. "Interest & Loan a/c s," an account book at the Office of the Scully Estates, Lincoln, Ill.; letter, Scully to John Williams & Co., May 20, 1870, Black and Williams Papers.
4. Lincoln *Weekly Herald,* April 20 and August 24, 1859.
5. Lincoln *Herald,* October 28, 1869.
6. Ibid., February 17, 1870.
7. William J. Stewart, "Speculation and Nebraska's Public Domain, 1863–1872," *Nebraska History* 45 (September, 1964): 265.
8. "1865–1870, Beatrice, Brownsville, Nebr., Cash 674 to 3773," General Land Office, National Archives, National Records Center, Suitland, Maryland. Although the totals are the same, the Scully entries in tract books 129 and 130 for the Beatrice Land Office show entries for June 14, which have been used in this text.
9. William McGalliard to John Williams and Company, March 1, 1871, Black and Williams Papers.
10. "1865–1870, Beatrice, Brownsville, Nebr., Cash 674 to 3773."
11. "Kan. Junction City, Cash Sales, Jan. 1, 1869 to March 31, 1871. 1074 to 2718," General Land Office, National Archives, National Records Center, Suitland, Maryland.
12. A. E. Case Collection, Spencer Research Library, University of Kansas, Lawrence.

13. Homer Edward Socolofsky, "The Scully Land System in Marion County," *Kansas Historical Quarterly* 18 (November, 1950): 340.
14. Letters to J. M. McFarland, December 14 and 19, 1876, A. E. Case Collection.
15. Yasuo Okada, *Public Lands and Pioneer Farmers: Gage County, Nebraska, 1850–1900* (Tokyo: Keio Economic Society, 1971), p. 73.
16. "Net income from American lands, after paying for improvements. . . ."
17. Ibid.
18. Lincoln *Herald*, November 13, 1873. On February 9, 1882, this newspaper reported that Parks was on the territorial supreme court bench in Wyoming.

CHAPTER 6

1. *History of Logan County, Illinois: Its Past and Present* (Chicago: Donnelley, Loyd & Co., 1878), p. 360; Lawrence B. Stringer, *History of Logan County* (Chicago: Pioneer Publ. Co., 1911), vol. 2, p. 364.
2. Among all of the stories about the Scully Estates that possess a gossipy quality or a legendary or rumor characteristic, the one concerning the paternity of John Scully by William is the most persistent. For example, Mark Hoblitt, Lincoln, Illinois, was eighty-three years of age when he told the author on August 30, 1963, that he remembered hearing that William Scully had acknowledged in court that he was the father of John Scully. There seems to be no such court record. Hoblitt remembered playing with John Scully's children; he also remembered having seen William Scully and having boxed with Scully's son Thomas. Legal adoption papers of John by William are not available either, although this adoption was mentioned to the author by Violet Scully (Mrs. Thomas A.) on November 14, 1962. William did provide extensive wealth to John's two children and perhaps some to John's wife.
3. Beaver, *William Scully*, p. 112, shows a lease issued to Henry Zumwalt for 80 acres for a five-year period, dating from July 20, 1867.
4. "In this Book are kept the Rent Accounts of the Tenants of Wm Scully in Logan and Tazewell Counties—Ills., commencing March 1/1870," Office of the Scully Estates, Lincoln, Ill.
5. "Hedge Book of William Scully's lands—Logan County," Office of the Scully Estates, Lincoln, Ill.
6. A. E. Case to Scully and Koehnle, December 21, 1876, Case Collection.
7. Case Collection.
8. Roger Andrew Winsor, "Artificial Drainage of East Central Illinois, 1820–1920" (Ph.D. diss., University of Illinois-Urbana, 1975).
9. "In this Book are kept the Rent Accounts," pp. 100–101, shows allowance to a tenant for half of the cost of cutting a ditch in 1870.
10. "Net income from American lands."
11. Beaver, *William Scully*, pp. 52–53.
12. Ibid., p. 53; interview with Nicholas Hamer, Hartsburg, Ill., September 1, 1963, and with Andy Gardner, Lincoln, Ill., September 1, 1963.
13. Interview with Nicholas Hamer, Hartsburg, Ill., September 1, 1963; Beaver, *William Scully*, p. 55.
14. Beaver, *William Scully*, pp. 53–60.
15. Ibid., p. 57.
16. Correspondence in the Office of the Scully Estates, Lincoln, Ill.

17. Ibid.
18. Beaver, *William Scully,* p. 56; undated clipping [ca. 1896] from the Kansas City *Star,* taken from the Chicago *Inter Ocean,* reported a $100,000 expenditure on tiling Scully lands in Illinois. Actually there are about 500 miles.
19. Beaver, *William Scully,* p. 54; James M. Stewart, "Farm Buildings: Leasing Arrangements [on] Scully Estates," *Proceedings,* Annual Winter Meeting, Illinois Society of Professional Farm Managers and Rural Appraisers (January 30–31, 1964), p. 3. Penalties were imposed for destruction or damage to the "drainage and erosion control system."
20. Interview with Mark Hoblitt, Lincoln, Ill., August 30, 1963.
21. Interview with J. C. McIntosh, Marion, Kans., April 5, 1947.
22. Lincoln *Herald,* June 8, 1882.
23. Notes in Office of the Scully Estates, Lincoln, Ill.
24. Lincoln *Herald,* April 9 and 16, 1885; Lincoln *Times,* April 9 and 16, 1885.
25. Lincoln *Herald,* December 17, 1885. This paper said that Trapp would take a position with Scully and Koehnle about January 1.
26. Beaver, *William Scully,* p. 51; interview with James M. Stewart, Lincoln, Ill., November 15, 1962.
27. Socolofský, "Scully Land System in Marion County," p. 347.
28. St. Louis *Post-Dispatch,* March 31, 1901; Gates, *Pioneer Landlords,* p. 37; Kansas City *Star,* January 26, 1919. This sale came about because of a favor granted to Scully by the governor.
29. A. E. Case to Scully and Koehnle, April 17, 1880, Case Collection.
30. "July 1/90 'Mr. Koehnle & I reckoned the Total cost of administration etc. of all my American Estates to be as follows for the next 3 to 5 years,'" memorandum in Office of the Scully Estates, Lincoln, Ill.
31. "Interest and Loan a/c s," Office of the Scully Estates, Lincoln, Ill. Early in the twentieth century the Lincoln bank of which Koehnle was president needed money quickly. Koehnle, as Scully's agent, bought $1.5 million of Russian bonds with Scully's money. The bonds had paid $600,000 interest by 1906. Thereafter they declined in value and were repudiated by the new Soviet government after 1917.
32. "Conditions of an eligible security," in Office of the Scully Estates, Lincoln, Ill.
33. St. Louis *Post-Dispatch,* March 31, 1901.
34. Index to Court Records, Logan County Circuit Court.
35. File 6858, box 303, Logan County Circuit Court; Lincoln *Times,* February 23, 1882.
36. Gates, *Pioneer Landlords,* p. 54.
37. *Hamilton* v. *Scully,* Supreme Court of Illinois, October 27, 1886. The Hamilton lands included 3,841 acres between Buffalo and Dawson and 300 acres west of Springfield. The 300 acres were sold prior to 1952.
38. St. Louis *Post-Dispatch,* March 31, 1901.
39. Beaver, *William Scully,* pp. 47–50.
40. Ibid.

CHAPTER 7

1. Office of State Papers, Dublin Castle, index to Ballycohey Outrage file.
2. Scully to Williams, November 30, 1870, Black and Williams Papers.

3. Frank W. Bill of the Bloomington *Daily Pantagraph,* in a letter to James M. Stewart, October 6, 1961, Office of the Scully Estates, Lincoln, Ill., sees the Rothamsted connection in the legume requirements in Scully leases. From early 1872 until early 1874, mail to William Scully was directed to "The Rectory, Hatfield, Herts [Hertfordshire], England," Black and Williams Papers.
4. "Address given by Thomas A. Scully at the annual meeting of the Logan County Historical Society, Emden, Illinois, February 24, 1957." Thomas Scully said that his father bought the buffalo bones "by hundreds of tons."
5. Interview with Violet Scully, November 20, 1962.
6. Certified copy of an entry of marriage, General Register Office, Somerset House, London.
7. "Will of Mrs. Enriqueta Angela Scully, June 4, 1931," copy in the Office of the Scully Estates, Lincoln, Ill.
8. Letter to the author from Violet Scully, January 17, 1972.
9. Ibid.
10. Lincoln *Herald,* May 14, 28, July 2, 9, 30, and September 3, 1885.
11. Kansas City *Star,* January 27, 1919.
12. "Some History of the Scully Family"; letter from A. I. H. Parry to Violet Scully, July 3, 1962, copy in Office of the Scully Estates, Lincoln, Ill.
13. Lincoln *Evening Courier,* April 1, 1936.
14. Kansas City *Star,* January 2, 1905.
15. St. Louis *Post-Dispatch,* March 31, 1901.
16. "The following seem to be the causes of the ruin of States," memorandum in Office of the Scully Estates, Lincoln, Ill.
17. Memorandum in Office of the Scully Estates, Lincoln, Ill.; brackets enclose editorial additions provided because of torn places in this manuscript.
18. St. Louis *Post-Dispatch,* March 31, 1901.
19. Kansas City *Star,* March 12, 1947.
20. St. Louis *Post-Dispatch,* March 31, 1901; letter to the author from Violet Scully, January 17, 1972.
21. St. Louis *Post-Dispatch,* March 31, 1901.
22. Kansas City *Star,* January 2, 1905.
23. "Return of Owners of Land of One Acre and Upwards in the Several Counties, Counties of Cities, and Counties of Towns in Ireland," *British Parliamentary Papers,* Session of 1876, vol. 80.

CHAPTER 8

1. Gates, *Pioneer Landlords,* p. 50, quoting the Pontiac *Free Trader,* December 20, 1878.
2. Chicago *Morning News,* May 5, 1887.
3. Gates, *Pioneer Landlords,* p. 50.
4. *Laws of Illinois, 1887,* p. 129.
5. *Nebraska Farmer,* July, 1880, p. 170, from the *Irish World.* The Marquis of Sligo, one of the great landowners of Ireland, possessed more than 100,000 acres with a yearly rental in 1883 of £19,000. The Baron Oranmore was another large Irish landholder. Michael Sullivant was a member of a family who acquired 53,000 acres in central Ohio with Revolutionary War military warrants. He bought about 80,000 acres in the Illinois counties of Champaign,

Ford, Livingston, and Vermillion from the government and the Illinois Central Railroad. His experimental operations at "Broadlands" and later at "Burr Oak" were termed early bonanza farms. Sullivant overextended himself financially and lost much of his huge estate as a consequence of the Panic of 1873.

6. Lincoln *Times*, March 25, 1880.
7. Chicago *Times*, March 30, 1880.
8. Lincoln *Times*, April 1, 1880.
9. Ibid., July 8, 1880.
10. *Republican* (———, Ill.), February 16, 1882.
11. Lincoln *Times*, March 15, 1883, had an article headlined, "Scully's Scalpers"; Gates, *Pioneer Landlords*, p. 55.
12. Gates, *Pioneer Landlords*, p. 53, quoting the Pontiac *Free Trader and Observer*, April 8, 1887.
13. Letter to the author from James M. Stewart, April 10, 1964.
14. Charlotte Clabaugh, "Scully Had Rough Time in Ireland before Coming to the United States," Superior (Nebraska) *Express*, April 6, 1967. Nebraska papers that were the most strongly anti-Scully were the Nelson *Gazette* and the *Gage County Democrat* of Beatrice.
15. Beaver, *William Scully*, p. 76.
16. Chicago *Morning News*, May 5, 1887.
17. Gates, *Pioneer Landlords*, p. 58. Illinois papers that were the most strongly anti-Scully include the Chicago *Tribune*, the Chicago *Inter Ocean*, the Bloomington *Pantagraph*, the Lincoln *Herald*, the Paxton *Record*, the Pontiac *Free Trader and Observer*, the Princeton *Republican*, the Pontiac *Sentinel*, the Piatt *Republican*, the Piatt *Independent* (Monticello), and the Paxton *Sentinel*.
18. Michael J. Brodhead, "The Early Career of E. W. Hoch, 1870–1904" (master's thesis, University of Kansas, 1962), p. 41.
19. Marion *Register*, May 5, 1886.
20. Ibid., January 12 to July 27, 1887.
21. Brodhead, "The Early Career of E. W. Hoch," pp. 40–41.
22. Marion *Register*, February 9 to April 27, 1887; Marion *Record*, February 25, 1887. Kansas newspapers that were most strongly anti-Scully included the Topeka *Commonwealth*, the Peabody *Graphic*, the Atchison *Champion*, the *Marshall County Democrat* of Marysville, and the nearby Kansas City *Star*, which circulated in Kansas.
23. *Senate Journal, Kansas, 1891*, pp. 120, 130, 134, 192, 594, 619, 768, and 774.
24. *Congressional Record*, 49 Cong., 1 sess., vol. 17, pt. 8, pp. 7830–34; *Statutes of the United States of America, Passed at the Second Session of the Forty-Ninth Congress, 1886–1887* (Washington, D.C.: Government Printing Office, 1887), pp. 476–77.
25. New York *Times*, January 26 and March 20, 1886.
26. New York *World*, July 14, 1888; Philadelphia *Evening Star*, August 28, 1888.
27. Memorandum in Office of the Scully Estates, Lincoln, Ill.
28. Gates, *Pioneer Landlords*, p. 60.
29. "Deed Record," Marion county, Kans., vol. 69, pp. 270, 279.
30. Letter from Oglesby to Scully, May 13, 1886, dictated copy in Oglesby Collection, Illinois State Historical Society, Springfield, Ill.
31. John Davis, "Alien Landlordism in America," in C. F. Taylor, ed., *The Land Question from Various Points of View* (Philadelphia: privately published, 1898), p. 57. Davis served in Congress from 1891 to 1895. His district included

Dickinson and Marshall counties, but not the other Kansas counties where Scully had land.

32. Gates, *Pioneer Landlords,* p. 48. Gates is talking about landlords in general in this passage.
33. Brodhead, "The Early Career of E. W. Hoch," p. 44.
34. Chicago *Morning News,* May 5, 1887.
35. Gates, *Pioneer Landlords,* p. 50, quoting the Pontiac *Free Trader,* December 20, 1878; Beaver, *William Scully,* pp. 69–70, quoting the Bloomington *Weekly Pantagraph,* February 11, 1876, and March 25, 1887; Chicago *Tribune,* March 11, 1887, and October 19, 1906; November, 1900, issue of the Lincoln *Weekly Courier;* Marion *Record,* February 25, 1887.
36. Henry George, "More about American Landlordism," *North American Review* 142 (April, 1886): 398–99.
37. Chicago *Morning News,* May 5, 1887.
38. Gates, *Pioneer Landlords,* p. 54.
39. St. Louis *Post-Dispatch,* March 31, 1901.
40. Chicago *Morning News,* May 5, 1887.
41. Ibid.
42. St. Louis *Post-Dispatch,* March 31, 1901.
43. Davis, "Alien Landlordism in America," p. 59.
44. Chicago *Morning News,* May 5, 1887.
45. Kansas City *Star,* January 2, 1905; *History of Logan County* (Chicago, 1886), p. 366.
46. Gates, *Pioneer Landlords,* pp. 55–56.
47. Chicago *Morning News,* May 5, 1887.
48. St. Louis *Post-Dispatch,* March 31, 1901.
49. Okada, *Public Lands,* pp. 134–35.
50. Chicago *Morning News,* May 5, 1887.
51. Ibid.
52. Gates, *Pioneer Landlords,* pp. 59–60.

CHAPTER 9

1. Gates, *Pioneer Landlords,* p. 46.
2. "Miscellaneous Record," Marion County, Kans., Register of Deeds, vol. 4, pp. 75–79.
3. Lease register, Logan County, Office of the Scully Estates, Lincoln, Ill. In 1901 Scully said that his Logan County land rented for $4 per acre, probably meaning that his net income was that amount.
4. St. Louis *Post-Dispatch,* March 31, 1901.
5. Leslie Gillette, to Koehnle and Trapp, January 2, 1893, Office of the Scully Estates, Beatrice, Nebr.
6. Ibid., September 11 and November 29, 1893.
7. Ibid., dated March, with no year, probably 1894, with no indication of Koehnle and Trapp's response to the request to provide financial aid to buy corn for tenants.
8. A. M. F. Randolph, *Kansas Reports,* vol. 40, p. 396.
9. Kansas City *Times,* November 6, 1946; St. Louis *Post-Dispatch,* March 31, 1901. A variation of this list also had "he must not have petty lawsuits," to

which a newspaper reporter added: "The farmers in the old tale who exhausted their lands in litigation over a bull calf were not lessees on Scully lands."

10. Kansas City *Star*, September 15, 1899; New York *Times*, October 14, 1895.
11. Kansas City *Star*, October 15, 1899, and April 6, 1901.
12. Ibid., September 15, 1899. This paper said this land was being purchased for Scully's son and widowed daughter-in-law (presumably Mrs. Louise C. Scully), whereas the New York *Times* said that Mrs. William Scully was buying the Bates County land. According to the Kansas City *Star*, Scully "paid about 1½ million dollars" for the Bates County land, probably a little high for the depression years of 1894–96, but much closer to reality than the one hundred to two hundred thousand dollars suggested by Gates, *Pioneer Landlords*, p. 43, or the Kansas City *Times*, November 6, 1946.
13. Prior to the Great Chicago Fire of 1871, Scully had purchased and sold a few Chicago town lots, so he had owned some Cook County property at one time.
14. Kansas City *Star*, September 28, 1895; copy of Scully's naturalization proceedings, filed at the Supreme Court of the District of Columbia on October 17, 1900, and affirmed in testimony dated December 12, 1907, Office of the Scully Estates, Lincoln, Ill.
15. St. Louis *Post-Dispatch*, March 31, 1901; a letter to the author from James M. Stewart, January 23, 1963, says that the younger William died in southern France.
16. Chicago *Morning News*, May 5, 1887.
17. Telephoned information from the keeper of local collections, Kensington and Chelsea Public Library, London, December 1, 1971. Mrs. E. Angela Scully's name was on the voter's list in 1905 and later.
18. J. Ed C. Fisher, agent at Beatrice and earlier a close friend of the younger William Scully's, wrote to Koehnle and Trapp on May 8, 1899, that "I take out my first Citizen Papers today as per request of Mr. Scully."
19. New York *Times*, November 16, 1895. In a similar story the Kansas City *Star*, undated clipping, ca. 1895, from the Chicago *Inter Ocean*, said: "Before leaving England, a few weeks ago, Mr. Scully sold every inch of English property and all in Ireland not entailed, and has now only two tenants and a little grazing land."
20. St. Louis *Post-Dispatch*, March 31, 1901; Kansas City *Star*, September 28, 1895. Scully was also listed as a member of the aristocracy in an article on alien landlords in the New York *Recorder*, reprinted in the Kansas City *Star*, February 9, 1895.
21. St. Louis *Post-Dispatch*, March 31, 1901; Kansas City *Star*, January 2, 1905.
22. Interview with James M. Stewart, November 20, 1962.
23. Ledgers in the Office of the Scully Estates, Lincoln, Ill. Mrs. E. H. Gurley paid the taxes due on the house to date of purchase. General Alger's rent was $6,000 per year. Thomas Scully, in "Some History of the Scully Family," described the Gurley House as "a very poorly architected house."
24. "Record Book," Office of the Scully Estates, Lincoln, Ill.
25. Chicago *Morning News*, May 5, 1887.
26. "Some History of the Scully Family"; Remarks of T. A. Scully to J. M. Stewart, June 9, 1958.
27. Note in William Scully's handwriting, dated 1900, Office of the Scully Estates, Lincoln, Ill.

28. Interview with J. M. Stewart, November 20, 1962; Lincoln *Evening Courier*, April 1, 1936.
29. Brodhead, "The Early Career of E. W. Hoch," p. 44. Editor Hoch's November 1, 1895, editorial was premature about actual citizenship for William Scully.
30. Lincoln *Daily Courier*, November 1, 1895; Fox to Koehnle and Trapp, April 21, 1896.
31. Brodhead, "The Early Career of E. W. Hoch," p. 44.
32. Interview with Mark Hoblit, August 30, 1963.
33. Kansas City *Times*, November 6, 1946.
34. Nuckolls County expense statement for June, 1902.
35. St. Louis *Post-Dispatch*, March 31, 1901.
36. Kansas City *Star*, October 19, 1906.
37. Notes in the Office of the Scully Estates, Lincoln, Ill.
38. Interview with James M. Stewart, November 14, 1962. William Scully, according to the transfer deed of the Butler County land, was "temporarily in England" when the document was witnessed by Richard Westacott, vice-consul general of the United States, at London. John C. Scully, in his own will of June 3, 1952, said that the Butler County property totaled 9,030 acres.
39. "Will of William Scully," Office of Register of Wills, District of Columbia.
40. New York *Times*, August 26 and October 19, 1906; Kansas City *Star*, October 19, 1906.
41. Certified Copy of Entry of Death, "William Scully," Given at the General Register Office, Somerset House, London.
42. Kensal Green Cemetery, grave site #41012, section 153, row 3.

CHAPTER 10

1. St. Louis *Post-Dispatch*, March 31, 1901.
2. Kansas City *Times*, November 6, 1946.
3. St. Louis *Post-Dispatch*, March 31, 1901; New York *Tribune*, October 19, 1906; Superior (Nebraska) *Express*, April 6, 1967.
4. Copy of agreement of January 31, 1907, in no. 14,065, administrative docket 36, "In re the estate of Wm Scully decd," in the Supreme Court of the District of Columbia.
5. Letter from E. Angela Scully to Trapp, March 29, 1909, Office of the Scully Estates, Lincoln, Ill.
6. "Some History of the Scully family"; copy of E. Angela Scully's English will, dated July 19, 1912; copy of a letter from Trapp to Mrs. Parry, October 2, 1915, all in Office of the Scully Estates, Lincoln, Ill. From the time of her operations in 1915 until her death in 1932 she employed a "good trained nurse" who lived with her.
7. "Some History of the Scully family"; copy of E. Angela Scully's English will.
8. Copy of E. Angela Scully's English will.
9. Copy of letters from Trapp to E. Angela Scully, June 5, 1913, and May 5, 1914; interview with James M. Stewart, November 14, 1962.
10. Copy of deposits to Mrs. E. Angela Scully's account, Office of the Scully Estates, Lincoln, Ill.
11. Interview with James M. Stewart, November 20, 1962.
12. "Probate proceedings of estate of John C. Scully," in abstract of title of the John

C. Scully lands from the F. S. Allen Abstract Company, El Dorado, Kans.; Kansas City *Star,* January 24, 1919. The Butler County oil strike is dealt with in Francis W. Schruben, *Wea Creek to El Dorado: Oil in Kansas, 1860–1920* (Columbia: University of Missouri Press, 1972).

13. Beaver, *William Scully,* p. 88, from the Decatur (Illinois) *Sunday Herald and Review,* August 10, 1941.
14. Remarks of T. A. Scully, June 9, 1958, to J. M. Stewart; "Some History of the Scully family."
15. Beaver, *William Scully,* p. 87.
16. Interview with Mark Hoblit, August 30, 1963. Pickrell had been Trapp's partner since Koehnle's death. Fox had long been the agent at Dwight, Ill.
17. Kansas City *Star,* January 26, 1919.
18. "The Scully Estate—Lords of 211,000 Acres," *Prairie Farmer* 91 (March 22, 1919): 516, 602–3; Kansas City *Star,* January 26, 1919.
19. Kansas City *Star,* January 26, 1919.
20. Interview with Mark Hoblit, August 30, 1963.
21. Kansas City *Star,* January 26, 1919.
22. Ibid., July 3, 1919.
23. Ibid., January 24 and 26, 1919.
24. Ibid., September 4, 1921.
25. Letters from John Powers to Trapp and Fox, September 22 and October 26, 1921, Office of the Scully Estates, Beatrice, Nebr.
26. *Kansas Reports,* vol. 120, pp. 638–39; Kansas City *Star,* January 22, 1923.
27. Letter from Powers to Trapp, May 26, 1924, Office of the Scully Estates, Beatrice, Nebr.
28. *Kansas Reports,* vol. 120, pp. 643–44; Kansas City *Star,* March 11, 1923.
29. "Will of Mrs. Enriqueta Angela Scully, June 4, 1931, Last American Will and Testament," copy in Office of the Scully Estates, Lincoln, Ill. This will reserved £40,000 sterling from Thomas's share of her estate to go to his first wife and children. On July 16, 1926, Isabell Burrell Blackwood, formerly the wife of Thomas Scully, signed a release from rights to real estate owned by Thomas. See Office of the Logan County Recorder, bk. 106, p. 151.
30. Russell L. Berry, *The Scully Estate and Its Cash-leasing System in the Midwest* (Brookings: South Dakota Agricultural Experiment Station, 1966), p. 144.
31. Copy of letter from T. A. and Frederick Scully to Trapp, December 12, 1927, Office of the Scully Estates, Lincoln, Ill.
32. Certified copy of an entry of death, "Enrequita Angelita [*sic*] Lascurine Scully," given at the General Register Office, Somerset House, London. "Gall stones—gall bladder removed" was also listed under the cause of death, and no physician was in attendance. Angela's body was buried next to William's in Kensal Green Cemetery. The quotation under her birth and death dates was: "Be Thou Faithful unto death, and I will give thee a crown of life."
33. "Will of Mrs. Enriqueta Angela Scully, June 4, 1931."
34. Letter from Fisher to Trapp, January 10, 1907, Office of the Scully Estates, Beatrice, Nebr.
35. Kansas City *Star,* January 26, 1919.
36. Copy of letter from Trapp to Frederick Scully, November 8, 1932, Office of the Scully Estates, Lincoln, Ill.
37. Lincoln *Evening Courier,* April 1, 1936; "Some History of the Scully Family."

38. Correspondence in files of the Office of the Scully Estates, Lincoln, Ill.; Kansas City *Times*, July 30, 1941.
39. Kansas City *Star*, March 12, 1947; Hutchinson (Kans.) *News-Herald*, March 13, 1947.
40. Berry, *Scully Estate*, p. 148.
41. C. M. Harger, "Biggest Farm Landlord," *Country Gentleman*, vol. 84, no. 25 (June 21, 1919), pp. 3–4, 26, 28; Boyd Rist, "Our Biggest Farm Landlord," *Farm Journal*, September, 1925, pp. 14, 89; Boyd Rist, "Legumes Pay in Tenant Landlord Farming," *Breeders Gazette* 90 (August 26, 1926): 169; *General Statutes of Kansas, 1935*, 67-31 to 67-533. The index in this volume lists "Scully lease," but Scully is not mentioned in the law.
42. Alfred H. Sinks, "Making Good Land Better," *Country Gentleman*, vol. 120, no. 4 (April, 1950), pp. 36, 132–35.
43. Berry, *Scully Estate*, p. 156.
44. Ibid., p. 186.
45. Ibid., p. 206.
46. Ibid., pp. 216–24.
47. Socolofsky, "Scully Land System in Marion County," p. 359; an unidentified Washington newspaper clipping dated October 30, 1942.
48. "Papers in the Matter of the Estate of Frederick Scully, Deceased," in Marion County probate court.
49. Lincoln (Nebr.) *Journal*, May 2, 1954, January 18, April 3, and June 21, 1955; Lincoln (Nebr.) *Star*, June 19, 1954, January 19, June 21, and August 20, 1955.
50. Stewart, "Farm Buildings," p. 5.
51. Ibid.
52. Decatur (Ill.) *Herald*, July 13, 1961; Bloomington (Ill.) *Pantagraph*, July 13, 1961; Chicago *Sun-Times*, July 13, 1961; *Illinois State Journal* (Springfield), July 13, 1961.
53. Beaver, *William Scully*, p. 96.
54. Stewart, "Farm Buildings," p. 5.
55. Beaver, *William Scully*, pp. 95, 98. The children of Thomas's first marriage were left out of his will; they initiated a contest that was settled out of court.
56. Stewart, "Farm Buildings," p. 3; interview with James M. Stewart, March 18, 1977.
57. Richard Carter, "Scully Estates," *Proceedings*, Annual Winter Meeting, Illinois Society Professional Farm Managers and Rural Appraisers (January 30-31, 1964), p. 5.
58. Berry, *Scully Estates*, pp. 165, 181.
59. Ibid., p. 173.
60. Stewart, "Farm Buildings," p. 5.
61. Interview with D. W. Montgomery, August 22, 1977.
62. Dorothea Kahn Jaffe, "Incorporation, Partnership: New Tools for Owner-Farmer," *Christian Science Monitor*, eastern ed., May 25, 1961, sect. 2, p. 3.

Index

wswswswswswsws